THE WAY THINGS WORK

THE WAY THINGS WORK

AWARD PUBLICATIONS LIMITED

ISBN 0-86163-526-4

Copyright © 1989 Gruppo Editoriale Fabbri, Bompiani,
Sonzogno, Etas, S.p.A., Milan

First published in Great Britain 1991
by Award Publications Limited,
Spring House, Spring Place,
Kentish Town, London NW5 3BH

Text: Studio in/out

Translation: Susan Meadows

Picture credits: Giorgio Alisi, Andrea Corbella, Michelangelo
Miani, Ferdinando Russo, Mario Russo, Logical Studio Com-
munication, Studio Erre A 70, Studio Pitre, Studio Prisma,
Studio Sun-Fog, Studio U.T.S., Triagono Illustrazione

Production: Dima

Printed in Italy by Gruppo Editoriale Fabbri S.p.A., Milan

INSTRUMENTS FOR MEASUREMENT AND OBSERVATION

TECHNOLOGY AND AGRICULTURE

DOMESTIC TECHNOLOGY

COMMUNICATION AND INFORMATION

TRANSPORTATION

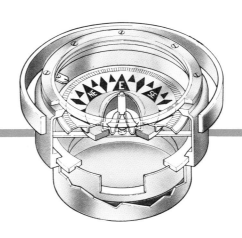

INSTRUMENTS
FOR MEASUREMENT
AND OBSERVATION

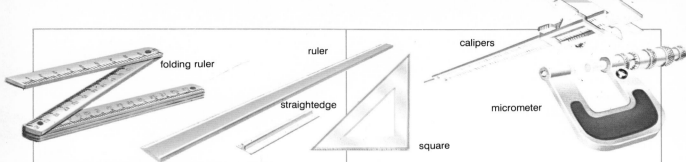

folding ruler

ruler

straightedge

square

calipers

micrometer

MEASUREMENT

What does "measuring" mean?
It is the comparison of something to a model, which is "a unit of measure." To find out, for example, how long a table is, we measure it with a yardstick, the correct unit

in this case, and see how many times a yard "goes" into the length of the table. Often we have to use sub-units to get the precise measurement. For the table, it might be feet or inches, corresponding to thirds or thirty-sixths of a yard.

watch

alarm clock

digital watch

stopwatch

digital clock

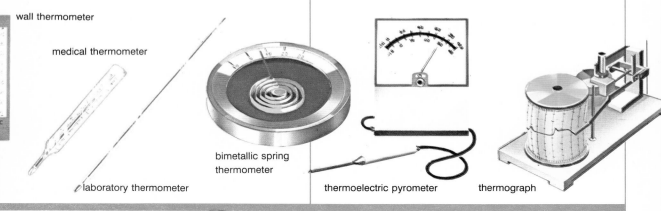

wall thermometer

medical thermometer

laboratory thermometer

bimetallic spring
thermometer

thermoelectric pyrometer

thermograph

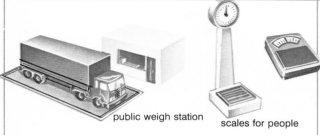

public weigh station

scales for people

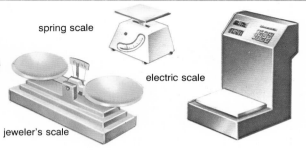

spring scale

electric scale

jeweler's scale

What can you measure?
There are many quantities that can be measured. A large part of scientific progress depends upon the ability to find a way to measure an existing quantity, or to define

a new one. The pictures on these two pages show instruments that measure many quantities, from length to time, from temperature to pressure and from weight to electric current.

Which instruments are used to measure length?

Some are found on the left: for example, a carpenter's folding ruler, an ordinary ruler, a straightedge, a square, calipers, which can measure thickness with a precision of 0.05 of a millimeter and a micrometer, which can measure thickness with a precision of 0.001 of a millimeter.

Is time only measured by clocks?

The instruments used for measuring time are called clocks. But there are many kinds of clocks which differ from each other, not so much in shape as in accuracy. On the opposite page you can see spring clocks (an alarm, a wristwatch, a stopwatch), a digital watch and a digital clock. There are also quartz clocks, for very accurate laboratory measurements, which can register up to one thousandth of a second.

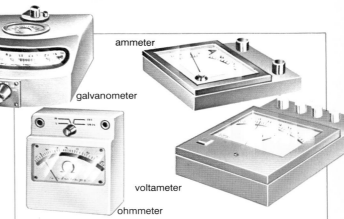

ammeter

galvanometer

voltameter

ohmmeter

Which instruments are used for measuring electricity?

Many electrical quantities can be measured. The galvanometer and ammeter measure the intensity of the current (the first is more sensitive). The voltameter measures the voltage and the ohmmeter, the resistance. The tester is an all-purpose instrument which measures different quantities depending on how it is connected to the circuit.

Are there many kinds of scales?

Yes, and they differ from one another in sensitivity. Pictured on the opposite page are scales for people, jeweler's scales, spring scales, and electrical scales with a display (often used in grocery stores). There are also precision scales for laboratories which can measure just a few grams. There is even the public weigh station which can hold very heavy loads but cannot distinguish grams.

Which instruments measure temperature?

Usually thermometers do this. Pictured on the opposite page are a wall thermometer, a medical thermometer, (for measuring body temperature), a very accurate laboratory thermometer, which can distinguish tenths of a degree, and a bimetallic spring thermometer, which can also use mercury, like the other three thermometers. Those containing mercury are more accurate, but are less sturdy. The thermoelectric pyrometer is used to measure temperatures over 5,500 degrees Fahrenheit (3000 °C), while the thermograph notes temperature change.

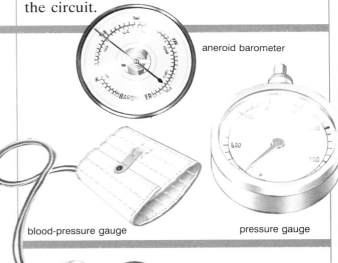

aneroid barometer

blood-pressure gauge

pressure gauge

Which instruments are used for measuring pressure?

Barometers are used for measuring atmospheric pressure. Those containing mercury are more sensitive than the aneroid ones, shown above. The pressure gauge measures the pressure of fluids compressed in a closed container such as air pressure in a tire. The blood-pressure gauge is used by doctors to measure the pressure of the blood.

THE COMPASS

Is the compass an instrument of measurement?

No, the compass doesn't measure anything. It is an indicator, an instrument used for orientation and navigation. The needle of a compass points to the north, from which you can find the other cardinal points and orient yourself in space.

Why does the needle point to the north?

The needle is made of a magnetic material which points in the same direction as the magnetic field of the earth, north. All magnetic objects have a south pole and a north pole. Poles that are alike repel each other; those that are different attract each other. The part of the needle that turns toward the north pole, then, is the needle's south pole.

A bar magnet enclosed in a sphere generates a field around itself similar to that of the earth.

Are there many magnetic materials?

Ancient people knew about magnetite, a mineral of iron which is naturally magnetic. There are other magnetic minerals, and many objects can become magnetic, for example, by being rubbed. Magnetism is connected to electricity. In natural magnets this is due to the electrical properties of the particles of the material itself. In other materials, it appears only in particular conditions, for example, when there is a specific temperature. Temperature can, in fact, change the electrical properties of materials.

angle of compass bearing in degrees

magnetized needle

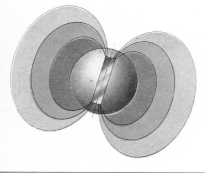

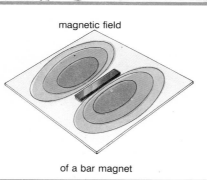

magnetic field

of a bar magnet

What is a magnetic field?

If you have a magnet, it will attract pieces of iron. Its action extends for a certain distance. The area in which it acts is its magnetic field. Its attraction is along "lines of force." The form of the magnetic field and the lines of force (which you feel but can't see) depend on the shape of the magnet. In the picture above, the field of a bar magnet is shown.

Can magnetic fields be produced with electricity?

Yes, electricity running through a wire always generates a magnetic field around the wire. If you put a compass near a wire in which there is electricity, the needle of the compass will deviate. The magnetic field generated by the current, being near, is stronger than that of the earth and prevails, modifying the behavior of the compass. The compass can also deviate when it is used near any other magnetic material or electrical installation.

Why does the earth have an electric field?

In the center of our planet there is an incandescent ball of molten rock which moves in currents similar to those of the rivers or seas. This incandescent ball turns at a speed different from that of the earth's crust above it. For reasons not yet understood, this difference in rotation creates a magnetic field. Like all other magnetic fields, this one has a north pole and a south pole, which are similar to, but not exactly the same as, the geographic poles.

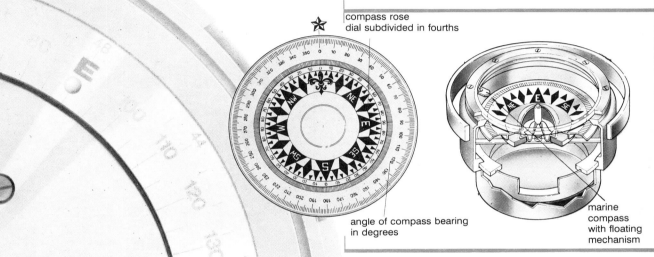

compass rose
dial subdivided in fourths

angle of compass bearing
in degrees

marine
compass
with floating
mechanism

If the magnetic and geographic poles are not the same, does a compass really point north?

The compass is oriented towards the magnetic north pole and, depending on the location, the difference in respect to the geographic north (which is called declination) may be perceptible. But declination charts have been made which correct the compass readings and give the right direction.

Are there many kinds of compasses?

The one that you see in the picture on these two pages is a very simple classic compass. The two small pictures on this page show a more complex navigational compass, suspended in order to function even when the ship is moving. There are even more complex instruments, like the gyroscope, which can be used on airplanes, missiles and rockets.

IN THE LABORATORY

Does a chemist use many instruments for measurement and observation?
Yes, a chemist does. On the table in the picture you can see some of the simpler ones, like the precision scale and the microscope. The Bunsen burner, mortar, cruets, test tubes and similar objects are used by the chemist to prepare substances to analyze and study.

Why do chemists often put colorants in solutions?
Chemists often use indicators, substances which, when added to a solution in small quantities, assume (or give the solution itself) a particular color. These substances can determine the chemical characteristics of the solution.

What can chemists discover by doing this?
For example, one of the most common tinctures, litmus (which comes from some species of lichen), turns blue in an alkaline solution and red in an acid solution. If you put litmus paper in a solution, you can tell from its color whether the solution is alkaline or acid. Other indicators can show whether there are, for example, traces of iodine or other particular substances, present.

How does a microscope work?

A microscope is an instrument of observation which allows you to enlarge the image of objects too small to be seen with the naked eye. Its operation is based on a system of lenses, that is, of objects which collect and converge light rays. In the tube of the microscope, there are more magnifying lenses. The closest to the object enlarges it, the next enlarges the image from the first, the third further enlarges the image from the second and so on. This system of lenses constitutes the objective. At the other end of the tube, the eyepiece allows you to look at the magnified image.

Why do magnified objects have to be very thin?

The microscope has a light source under the stage on which the specimen is placed. The specimen must be transparent to allow the light to pass through and to give a clear image.

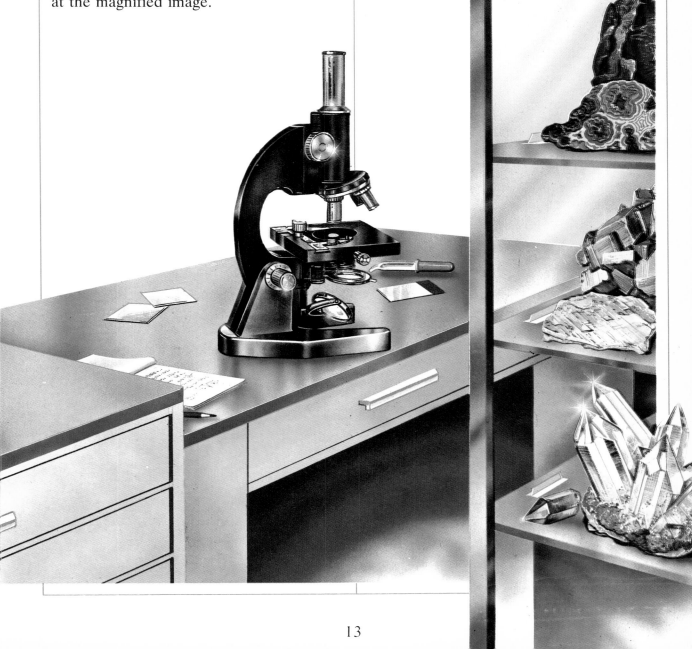

WEIGHT, PRICE AND RECEIPT

Are the new scales that you see in grocery stores electronic?

The weight is determined mechanically but the weighing generates an electric signal which is sent to a small electronic calculator in the machine. The electronic calculator visualizes the weight using a liquid crystal display and is able to calculate the price of the goods. If, using a numeric keyboard, you register the price of the merchandise per pound, the calculator supplies the product of that price times the weight, which is the total cost.

How is the weight found by these machines?

You can see in the picture on this page. The platform is lowered due to the force exercised by the object placed on it thus moving a vertical weight. By means of a balance beam, the vertical movement of the weight is transformed into the rotation of a pendulum connected to a counterweight. The pendulum is graduated and its scale is projected, with a system of lenses and mirrors, onto a screen where the result of the weighing can be read. Here the result is not usually in the form of digits (it is not given in numbers), but in analogical form (a needle that moves on the graduated scale in proportion to the movement of the pendulum which, in turn, is proportional to the degree that the platform is lowered).

Can these machines also print a receipt?

Yes, many can. The electronic calculator can be constructed with an incorporated printer for printing the cost of the merchandise on paper.

How is a cash register different from an electronic scale?

The cash register is not a machine for weighing. It is only used to make calculations, to produce a sales slip, a receipt or an invoice and to keep a record of all the operations done. It's not a measuring device, as the scale, electronic or not, is.

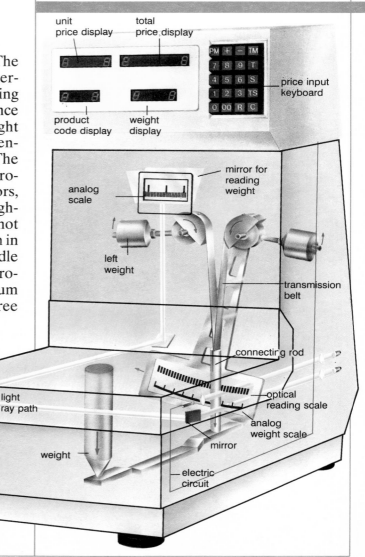

A modern scale which contains a small electronic calculator can give not only the weight of the item placed on it, but also the price whenever the cost of the unit, for example, pounds or kilograms, is registered.

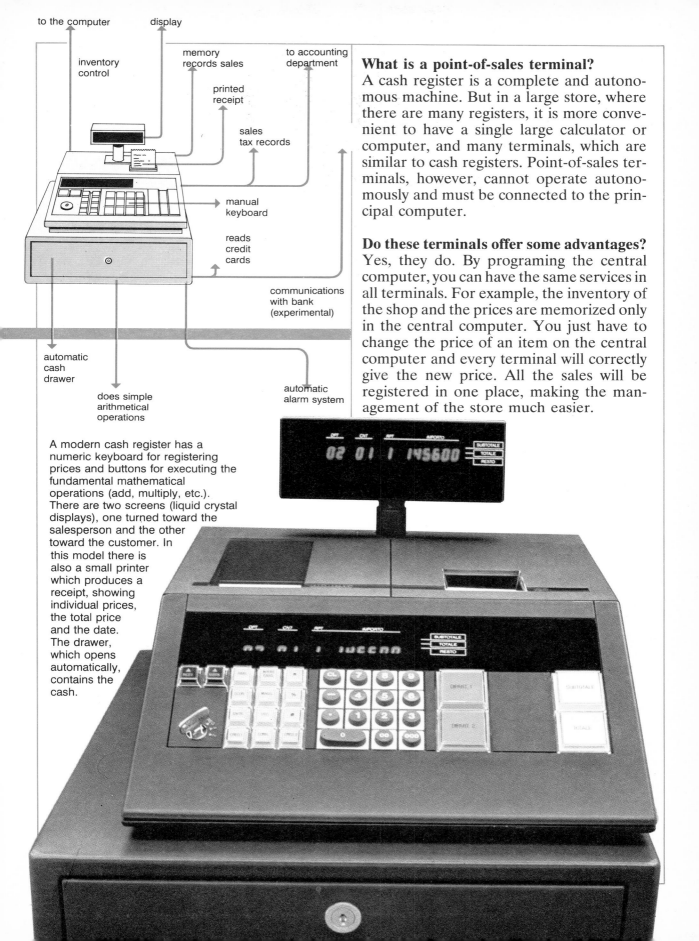

to the computer   display

inventory
control

memory
records sales   to accounting
department

printed
receipt

sales
tax records

manual
keyboard

reads
credit
cards

communications
with bank
(experimental)

automatic
cash
drawer

does simple
arithmetical
operations

automatic
alarm system

A modern cash register has a
numeric keyboard for registering
prices and buttons for executing the
fundamental mathematical
operations (add, multiply, etc.).
There are two screens (liquid crystal
displays), one turned toward the
salesperson and the other
toward the customer. In
this model there is
also a small printer
which produces a
receipt, showing
individual prices,
the total price
and the date.
The drawer,
which opens
automatically,
contains the
cash.

What is a point-of-sales terminal?

A cash register is a complete and autonomous machine. But in a large store, where there are many registers, it is more convenient to have a single large calculator or computer, and many terminals, which are similar to cash registers. Point-of-sales terminals, however, cannot operate autonomously and must be connected to the principal computer.

Do these terminals offer some advantages?

Yes, they do. By programing the central computer, you can have the same services in all terminals. For example, the inventory of the shop and the prices are memorized only in the central computer. You just have to change the price of an item on the central computer and every terminal will correctly give the new price. All the sales will be registered in one place, making the management of the store much easier.

ANGLES AND BINOCULARS

What is a protractor?

It's a very simple instrument, usually made of transparent plastic, for measuring angles on a piece of paper or a level area. It's a bit like a straightedge but has a circular shape. You match the zero with one side of the angle to be measured and then you read the value (usually in degrees) that corresponds to the other side of the angle.

What is a transit?

It is an instrument that measures angles. It is much more accurate than the protractor and can read horizontal and vertical angles. It has a telescope for distant reference points.

The transit is a portable instrument, small and light, with a telescope and a protractor inscribed on the glass for reading angles to the second of a degree.

measurement of the horizontal angle

When do you need to measure angles?

Other than on paper, for the preparation of plans (for machines, for example), they are used in topography for constructing houses, roads, tunnels or bridges.

Do the measurements have to be accurate?

For all technical work, yes, they do. For a bridge or a tunnel, for example, accuracy is fundamental. When you dig a tunnel, you usually decide in advance where it will come out and often the digging begins at both ends at the same time. The precise measurement of the angles of points that are very far from one another is indispensable in these cases. Otherwise the two ends of the tunnel would never meet. In these situations, the theodolite, or surveyor's transit, a more precise instrument than the protractor, is used.

How does the transit work?

The transit is essentially a telescope mounted on a tripod. It can rotate horizontally and can be raised or lowered. A level in the telescope makes it possible to establish the exact position of the horizontal plane and a plumb bob on the vertical axis of the telescope makes it possible to position the instrument precisely over the point where the measurement is to be taken. Each of the axes of the telescope runs at right angles through two graduated concentric disks, similar to protractors. To measure the horizontal angle between two lines, you put the instrument in position using the plumb bob. Then you sight through the telescope to a clearly visible point. You rotate the outer disk of the horizontal axis and set it at zero. Then you point the telescope at the second point and read the angle that is shown.

What is a telescope?

It is an optical instrument which is used to see distant objects. In its simplest form, it is a tube with two lenses at the ends. One, the objective, forms the real image of the object observed. The other, the ocular, is for seeing that image. The two lenses are placed in a way that enlarges the resulting image. Therefore, the object seems much closer. The image produced is inverted in an astronomical telescope. A third lens is required to see the image correctly. In modern telescopes, the simple lenses are replaced by complex systems, eliminating defects.

What are binoculars?

Binoculars are a pair of telescopes placed next to each other. Inside there is a particular system of prisms which makes it possible to use a much shorter tube than the traditional telescope. The two eyepieces produce an image that is much closer than that seen by the eyes. The effect of depth is produced by using both eyes.

How are the prisms of binoculars made?

They have the shape of half a cube, cut along the diagonal. They are paired in each of the two tubes of the binoculars in order to lengthen the path of the light rays (which allows for a shorter tube) and to give a straight image.

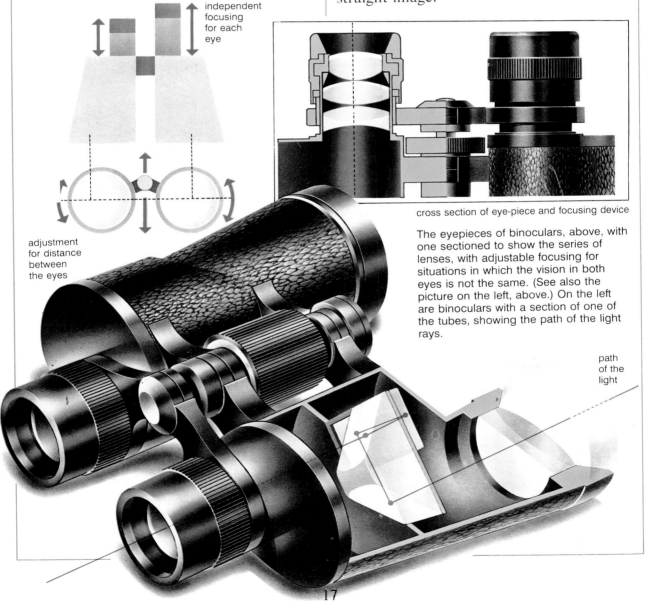

independent focusing for each eye

adjustment for distance between the eyes

cross section of eye-piece and focusing device

The eyepieces of binoculars, above, with one sectioned to show the series of lenses, with adjustable focusing for situations in which the vision in both eyes is not the same. (See also the picture on the left, above.) On the left are binoculars with a section of one of the tubes, showing the path of the light rays.

path of the light

QUARTZ WATCHES

How is time measured?

To measure time, we use some type of physical system that produces regular oscillations of the same length, for example, a pendulum. Once the length of each oscillation is established, time is measured by counting, or having some appropriate device count, the number of oscillations.

What mechanism is used in traditional watches?

In traditional watches the time standard is supplied by the oscillations of a small wheel, the balance wheel. A spiral spring is attached to the balance wheel, which tends to maintain it in a position of equilibrium. If the balance wheel receives a little push, it tends to oscillate regularly through the action of the spring.

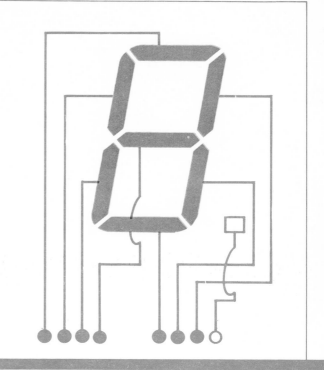

The electronic circuits which make up a digital watch can be much smaller than the set of wheels which constitute the fundamental mechanism of a traditional watch. For this reason, digital watches can have many different shapes and can include other functions. Here you can see a watch planned for a person who enjoys outdoor sports. Besides the time, it has a chronometer, barometer, altimeter and depth-gauge.

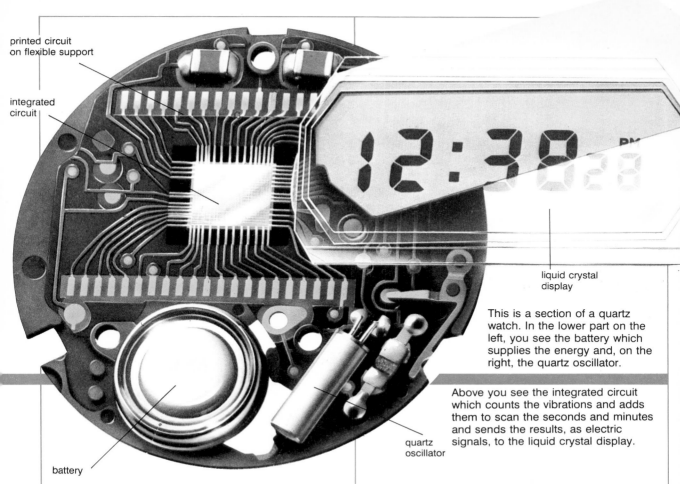

printed circuit
on flexible support

integrated
circuit

liquid crystal
display

This is a section of a quartz watch. In the lower part on the left, you see the battery which supplies the energy and, on the right, the quartz oscillator.

Above you see the integrated circuit which counts the vibrations and adds them to scan the seconds and minutes and sends the results, as electric signals, to the liquid crystal display.

quartz
oscillator

battery

What is the basic mechanism of a quartz watch?

If you take a thin sheet of correctly-cut quartz and give electrical impulses to the opposite sides, the sheet will become deformed, as if it had been compressed and stretched. These deformations, under particular conditions, take on a rhythmic modulation which have a precise duration, for example, one-hundredth of a second.

Does this phenomenon have a name?

Yes, it is called piezoelectricity, and it also works in reverse. That is, when the quartz sheet is deformed, a tension is generated between the two opposite sides, similar to that between the two poles of an electric battery. If the sheet vibrates rhythmically, the tension between the two sides will also vary with the same rhythm. In this way, electric impulses are produced, all of which have exactly the same duration.

Is piezoelectricity what makes quartz watches possible?

Yes, it is. If the faces of the vibrating quartz sheet are connected to an electronic device which is able to count the electrical impulses emitted and visualize the number of impulses counted, you have a quartz watch. Since the duration of the oscillations is very short, the appropriate circuits in a quartz watch visually supply hours, minutes and seconds directly to the display.

Is the quartz sheet equivalent to the balance wheel?

The quartz sheet functions as the oscillating system, which, in traditional watches, is done by the balance wheel.

Are quartz watches very accurate?

Their accuracy is much greater than that of traditional watches. It is on the order of a second every thousand days.

THE TELESCOPE

What is the difference between an ordinary telescope and one used for astronomy?

Both are instruments for observing distant objects, but the ordinary telescope uses only the laws of light refraction while the astronomical telescope is also based on the property of reflection of concave mirrors.

What are the properties of concave mirrors?

Flat mirrors, which we ordinarily use, give direct images of the same dimensions as the objects they reflect. Curved mirrors, instead, show deformed images, that is, enlarged or reduced or sometimes upside down, of what they reflect. Concave mirrors, in particular, give enlarged images of near objects and reduced and inverted images of distant objects, as if they were concentrated in a small space. Convex mirrors, instead, always give straight and reduced images, putting a large visual field in a small space.

Are there different kinds of telescopes?

There are different types of reflecting telescopes which have different placements and combinations of mirrors. Below you see three of them which take their names from the scientists who invented them: the Newton, the Cassegrain and the Coudé.

Are telescope mirrors very large?

Yes, those in astronomical observatories may have a diameter of one yard. One of the largest and most famous, in the Mount Palomar Observatory in California (on the facing page), has a diameter of more than 5.5 yards (5 m) and weighs fifteen tons.

Why are they so large?

One of the functions of a telescope is to collect more light than the human eye. The lens of a telescope is like the pupil of an eye, but much larger. For this reason, it collects more light and concentrates it in an intense band of light rays small enough to enter the pupil. The Mount Palomar Telescope, for example, can collect 360,000 times more light than our eyes. The larger the telescope lens, the more light it collects and the clearer the image it projects.

But isn't the image given by a concave mirror small?

A concave mirror gives a small but precise image. You can enlarge it to the dimensions you want, in a telescope, by using an ocular lens, which works a little like a microscope. The image of Mars, which to the naked eye appears as a red point, can be enlarged to the point in which its details, such as craters and canals, can be seen separately.

In the Newton reflector (1) the light is collected by a concave mirror, A (primary mirror), and then reflected by a flat mirror, B (secondary), inclined at forty-five degrees, which deviates it laterally to the lens. In the Cassegrain reflector (2), the light is collected by mirror A, perforated in the center, which sends it to a concave mirror, B, which reflects it to the lens below, passing through the hole in mirror A. The Coudé reflector (3), other than the primary A and secondary B mirrors, has a third mirror, C, which revolves. To follow the movement of the sky, you only have to move this mirror rather than moving the entire telescope.

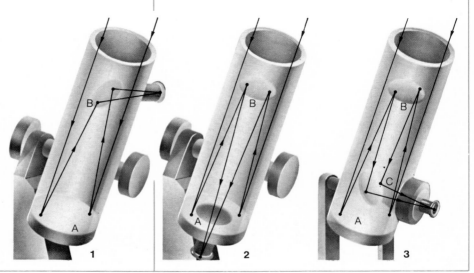

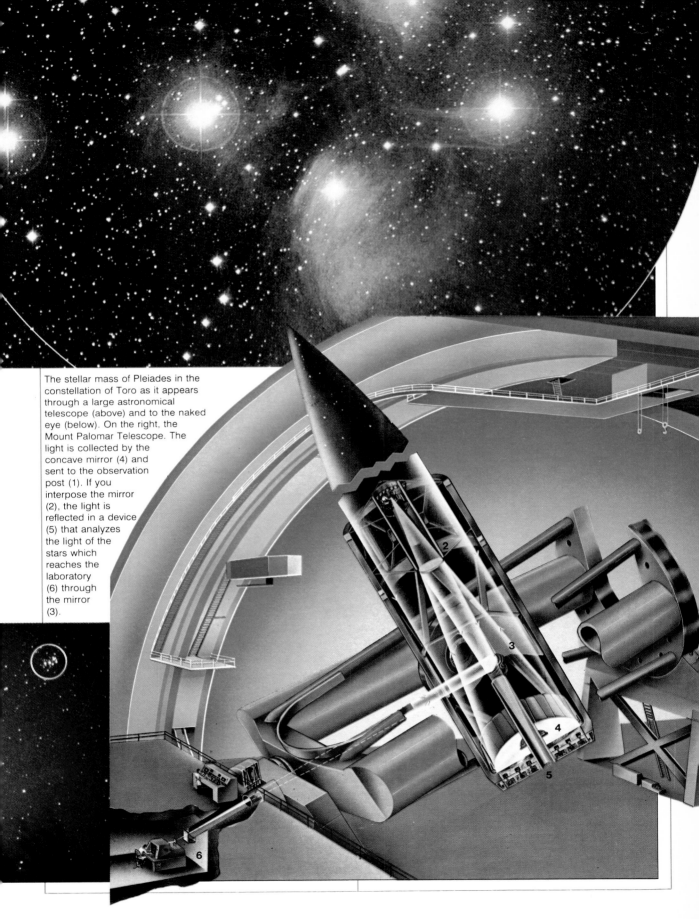

The stellar mass of Pleiades in the constellation of Toro as it appears through a large astronomical telescope (above) and to the naked eye (below). On the right, the Mount Palomar Telescope. The light is collected by the concave mirror (4) and sent to the observation post (1). If you interpose the mirror (2), the light is reflected in a device (5) that analyzes the light of the stars which reaches the laboratory (6) through the mirror (3).

RADAR AND SONAR

What is radar?

It is a device which can locate and identify moving objects by means of radio waves. Its name is an acronym: RAdio Detection And Range is the complete phrase.

How does it work?

The principle of radar is quite simple. Radio signals are sent by means of an antenna which are then reflected by a moving object, for example, an airplane or a ship. A receiving device, near the antenna, registers all the reflected signals. The analysis of these reflected signals gives information about the presence of the identified object, its distance and its velocity.

What are radio waves?

They are electromagnetic waves, like those of visible light, but with a much lower frequency and a much longer wavelength (more than a millimeter).

Is radar used often?

Nowadays radar is widely used. It is irreplaceable in cases where moving objects must be observed and identified, for military or civilian reasons, as in commercial airports.

How are objects revealed?

This visualization can occur directly by means of a light signal on a circular screen or a map representation of the area explored. In the second case, the system uses a computer which simulates the actual conditions of the space explored, showing a realistic representation on the video.

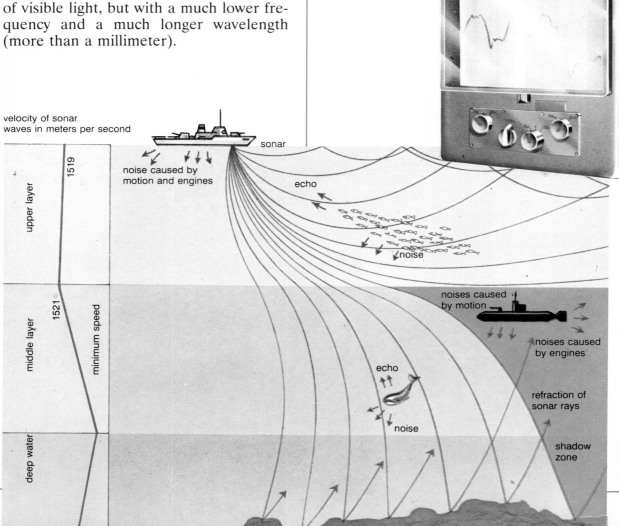

velocity of sonar waves in meters per second

upper layer — 1519

middle layer — 1521 — minimum speed

deep water

sonar

noise caused by motion and engines

echo

noise

noises caused by motion

noises caused by engines

echo

noise

refraction of sonar rays

shadow zone

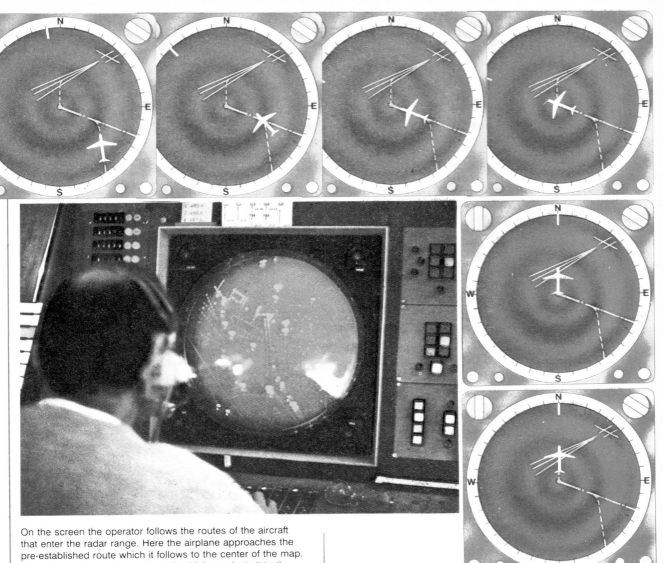

On the screen the operator follows the routes of the aircraft that enter the radar range. Here the airplane approaches the pre-established route which it follows to the center of the map. Then it veers toward the radio path which conducts it to the runway or to a new route.

Is sonar similar to radar?
The basic idea is similar, but in the case of sonar, we use sound waves instead of radio waves.

What are sound waves?
They are the variations in pressure that produce the sensation of sound.

Is the word "sonar" also an acronym?
Yes, it stands for SOund Navigation And Ranging; that is, acoustic navigation and location.

What sounds does sonar use?
It does not use sounds that we perceive, those typical of the voice or of musical instruments. It uses much higher frequency sounds called ultrasonics.

When is sonar used?
Sonar is used to locate objects in water, a means by which the radio waves of radar do not transmit as well as the sound waves of the ultrasonics. Sonar can detect the presence of submarines, large fish or any object capable of reflecting the waves emitted from the source.

Is sonar accurate?
It is not as accurate as radar. There may be echoes and interference, caused by the irregularity of the seabed or by variations in the density and temperature of the water.

REMOTE SURVEYING

What is remote surveying?
The words, in general, mean all operations of collecting and recording data about the earth which are done at a distance without direct contact with the region or the objects that are being surveyed. Usually remote surveys are done from airplanes or satellites.

Why are these surveys done at a distance?
Long-distance surveys of the land offer certain advantages over those done on land. The surveyors aboard airplanes and satellites can analyze hundreds of square miles in a few hours and can repeat the complete analysis of the same area a few days later. This allows for the observation of immense stretches, for example, very large forests which would be difficult to reach on foot.

What instruments are used?
Sensors which can register different types of radiation, from visible light to thermic emissions (infrared rays) and microwaves (radar waves), are used for remote surveys. They can also detect variations in the magnetic and gravitational fields of the earth.

Are photographs useful?
Taking photographs of the earth, in the area of the visible wavelengths, is very useful. Photographs of a particular zone can give a broad view of that area for use in making maps of the geologic characteristics and the man-made constructions. Even more useful is the multispectral photograph.

What is a multispectral photograph?
It is a photographic technique which consists of inserting special filters in the camera which record, on black-and-white film, different ranges of wavelengths of light on individual photograms. In this way, you can record the different components of the spectrum which give different information. Then you can combine the various shots taken at different wavelengths to construct high-resolution colored photograms.

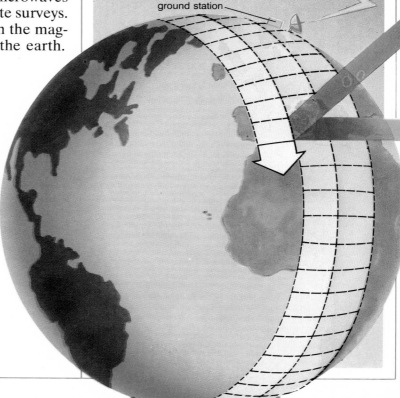

Information is transmitted to the station on earth when the satellite is directly in line with it. Otherwise the information is recorded aboard. On earth, the information is processed and stored on film or on a computer tape.

ground station

LANDSAT, an artificial satellite, revolves around the earth, flying over each point at the same time each day. The instruments aboard survey vertical bands and, in this way, gradually cover the whole earth like a ribbon wrapped around a ball. The instruments aboard LANDSAT electronically explore the area below, surveying the intensity of the sunlight reflected from the earth in four different "colors" or wavelengths. The four separate images are transformed into electronic impulses and transmitted to earth as numeric signals. They are received by a station in radio contact with the satellite. The map on the right represents the areas covered by these stations.

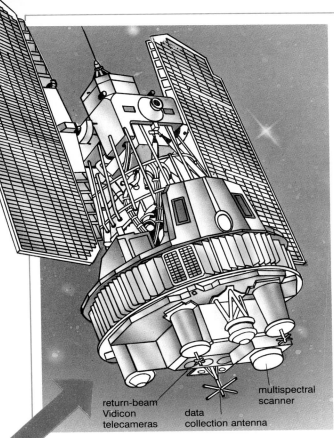

return-beam Vidicon telecameras

data collection antenna

multispectral scanner

Why do we survey infrared radiation?

All the substances on earth continuously emit infrared radiation (heat) and it is very useful to identify the differences in these natural thermic emissions. Geologists, for example, can identify particular minerals and discover even very small geological details this way.

Does infrared radiation also tell us something about the oceans?

Yes, the thermic maps of the oceans allow us to detect the presence of sea currents at different temperatures. They can also show the existence of underwater volcanic activity. They tell us which areas produce the most plankton (microscopic organisms that live in the sea), and they can show whether pollutants from industrial waste have been discharged into the sea.

What can we discover, at a distance, about the vegetation of an area?

Plants reflect a large amount of energy, in particular infrared energy. The wavelength range and the intensity of the radiation reflected are characteristics of each type of plant, its age, its health, and even the chemical composition of the soil. The analysis of infrared surveys, therefore, allows us to collect all of this information.

Is radar also used in telesurveys?

Yes, it is because the microwaves emitted by radar can penetrate clouds and make it possible to continuously observe the earth, day and night, in all atmospheric conditions. Microwaves can also penetrate vegetation to show the structure of the soil, even in regions covered by forests and jungles.

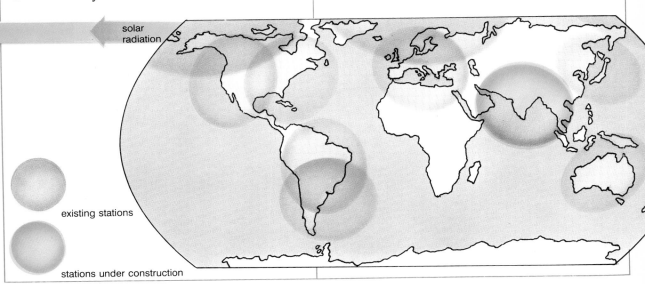

solar radiation

existing stations

stations under construction

Computerized Axial Tomography (CAT)

Are many instruments of measurement and observation used in medicine?

Yes, they are. Doctors must measure and analyze in order to judge the condition of their patients. The instruments used range from the simplest, like the stethoscope, which amplifies the sounds produced inside the human body, to very complex and refined instruments of observation, like x-ray techniques or computerized axial tomography, the equipment of which is shown on these two pages.

How does the x-ray work?

The x-ray is a photograph taken using x-rays instead of visible light. Unlike visible light, x-rays penetrate the body and pass through it to a degree that depends on the characteristics of the body itself. On special x-ray film, the internal human organs appear more or less transparent or opaque, depending upon their condition.

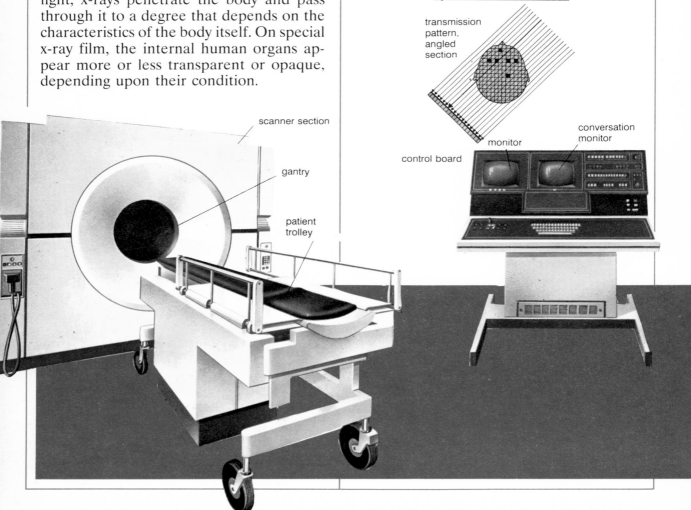

x-ray source (fan ray scan)

subject of the exam

movement of scanner

multiple detectors

transmission pattern, horizontal section

transmission pattern, angled section

scanner section

gantry

patient trolley

control board

monitor

conversation monitor

What is computerized axial tomography used for?

It is used to x-ray the human body in very small sections that may have a thickness of less than one millimeter. In this way, you can have a complete analysis of the torso, the limbs or the head, at any level. The traditional x-ray, instead, can only give an overall picture of a part of the organism.

How does CAT work?

As you see in the picture at the bottom of these pages, the patient is placed on a special trolley, which is movable and can be tilted. The trolley is inserted in varying degrees (depending on the part that is to be examined) into the tomography scanner unit which is equipped with an entry door or gantry. Around the circumference of the gantry is the x-ray system, made up of a crown of seven hundred twenty sensors which assure that the x-ray field covers the three hundred sixty degrees of the complete circumference. The x-ray source is a fanray scan.

How do you get a picture of the organism being examined?

Everything that is recorded by the sensors is transferred to a computer which processes the data and constructs the corresponding image. The computer is guided by a control panel from which an operator, using a monitor, can immediately verify the images and order necessary corrections in the placement of the equipment. On another panel is a diaphano scope, a special screen, litfrom behind, which allows for immediate viewing of the x-ray photographs.

Can you compare different x-rays by using the computer?

Yes, the computer controls all the tomography equipment by means of the control panel and with it, you can compare the photographs you have just made with those of a previous session, or with normal or pathological reports, filed in the magnetic memory of the computer. This makes the interpretation of the results of the tomography much easier and much quicker.

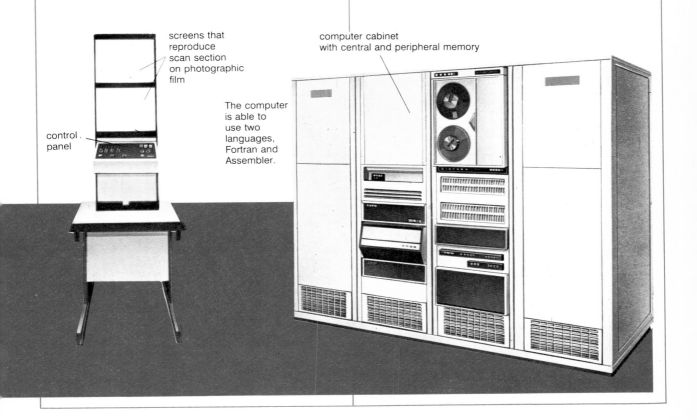

screens that reproduce scan section on photographic film

computer cabinet with central and peripheral memory

The computer is able to use two languages, Fortran and Assembler.

control panel

DID YOU KNOW...?

How does a stethoscope work?

The stethoscope is one of the simplest and oldest instruments of observation used in medicine. It helps the doctor to hear the sounds produced within the organism which, in their turn, give information about the patient's condition. One end of the instrument is placed on the body of the patient so as to collect all the sounds that come from the organism, excluding those from outside. The sounds reach the ears of the doctor by means of rubber tubes, positioned to exclude external noises.

What is an electrocardiogram?

It is the marks produced on paper by the electrocardiograph, an apparatus that registers the differences in potential of the electric activity of the heart within the area that surrounds the heart itself. The electrocardiogram does not give a direct image of the electrical activity of the heart but of the consequences that the activity has on the tissues around it.

What is an oscilloscope?

It is an electronic apparatus with which you can see at one time the course of variable electric quantities. The visualization comes about by means of a cathode ray tube, similar to the one used in televisions. The marks on the screen, produced by an electric beam, regularly move horizontally. Vertical variations in position are proportionate to the values which these quantities have at the time.

What is a sonometer?

It is a device used for studying the vibrations of a resonating string. The string is stretched over a soundbox which has support braces at both ends. The tension of the string can be varied by using weights. Another brace that can be moved on a graduated scale makes it possible to vary the portion of the string that can vibrate and, at the same time, to measure its length.

What is triangulation?

It is a system of measurement, introduced in 1617 by the Dutchman, Snellius, for measuring large surface areas. It is based on a geometric property of triangles which says that if you know the length of one side and the two angles which it makes with the other two sides, you can calculate the length of the two unknown sides. The procedure of triangulation consists of projecting a chain of triangles, each time using the known side of one triangle to calculate the successive one. With this system, instead of making many measurements of distances, you only have to make measurements of angles. This system is convenient because measuring angles is easier and gets more accurate results than measuring distances.

How can we study the inside of the earth?

The most immediate way to discover how the ground under us is constituted is to perforate it. Perforations are very much improved methods of excavation which allow us to reach depths of several thousands of yards, passing through very hard rock. For this operation, drilling derricks of varying shapes and dimensions, depending on the depth we want to reach, are prepared. The earth is perforated by using high-speed chisels made of very resistant material.

What about explosions?

If the surface of water is struck, waves are formed which move away from the center point in concentric circles. If you strike a piece of wood, waves spread inside it. The same happens in all materials. The vibrations produced by impacts or explosions are called "elastic waves" which are used for underground investigations because they travel at different speeds, depending on the material they pass through. After placing a series of receivers, you create an explosion in the ground or in the water. Depending on how long it takes the waves provoked by the impact to reach the receiving apparatus, you can establish what type of earth or rock these waves passed through.

TECHNOLOGY AND AGRICULTURE

PLOWS AND TRACTORS

What is plowing?
Plowing means preparing the earth for new cultivation, for successful planting. The earth is plowed by cutting, overturning and breaking up the turf. Working on the ground, turning it over and softening it, increases the circulation of water and air which helps the plants take root and grow.

Is the plow the tool that does this work?
Yes, the plow is the principal tillage tool. It breaks up the surface to a depth of six to sixteen inches (15-40 cm). It is a very old tool which, at one time, was powered by human muscle or strong animals. Modern plows are pulled by tractors and have from one to ten spades, called bottoms, attached to a frame. These spades have a variety of shapes and cutting edges, depending on the type of plowing to be done and the ground conditions. For deep plowing, the most common type is the moldboard which makes a furrow, overturning the soil on the surface.

What is the plowshare?
It is one of the three parts of the blade, specifically, the part that makes the furrow in the ground.

What are the other parts?
They are the moldboard and the landslide. The three parts are held together by a frame shaped like a three-sided wedge.

What are the moldboard and landslide used for?
The moldboard, placed above and behind the plowshare, turns over the soil, breaks it up and pushes it aside. The landslide, found behind the plowshare, slides along the bottom of the broken-up furrow and keeps the plow stable as it works.

What are plowshares made of?
They are made entirely of metal.

How much time does it take to plow the land?
The time needed to plow about 2.5 acres varies a great deal, depending on the equipment used, as you can see in the table on the opposite page. It takes from more than two days with a primitive plow pulled by two horses to a little more than half an hour with a gang plow pulled by a heavy tractor. The technique is always the same: the difference is in the energy used by the particular mechanism employed.

What is a tractor?
It is a particular type of vehicle suitable for pulling other machinery. It may pull other vehicles, farm machines or special equipment.

Are tractors used frequently in farming?
Yes, they are the most useful all-purpose agricultural machines. Running on gas or diesel, they act as power units and pulling devices for the more specialized seasonal equipment, like plows and planters.

Do tractors have closed cabins?
Not all do, but most modern tractors, like trucks, have closed cabins. They are sound-proof and shock-proof and sometimes even have air conditioning. Farmers no longer need to put up with bad weather and discomfort, but can concentrate on the work at hand.

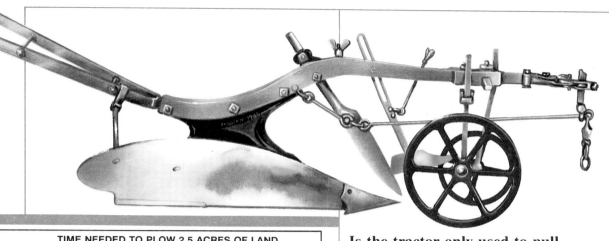

TIME NEEDED TO PLOW 2.5 ACRES OF LAND	
single-share plow and two horses	13 h. 42 min.
twin-share plow and four horses	5 h. 57 min.
twin-share plow and tractor	4 h. 19 min.
small gang plow and tractor	2 h. 7 min.
gang plow and heavy tractor	0 h. 40 min.

Is the tractor only used to pull other machines?

That is its principal function, but a tractor usually has several means of providing power to the implements attached to it. The hydraulic pump, for example, supplies power to lift and lower attached equipment, making it possible to automatically adjust heights and angles on harrows, plows and other equipment. With this system, farmers can set the depth to which they want to till the land. A power takeoff provides energy for other machines that are mounted on or pulled by the tractor. It, for example, supplies power to the moving parts of mowers, hay balers and spray pumps.

FARM MACHINES

Are there many kinds of farm machines?

Until this century, there were a few simple farm machines. But in the last decades, technology has made possible many different machines, specialized for different jobs and, therefore, able to do them in the most efficient way.

What is the machine shown below used for?

It is a very complicated machine, used to harvest tubers, in particular, potatoes. The soil is separated out using chain elevators and then a special device, the sorting elevator, is used to separate the tubers.

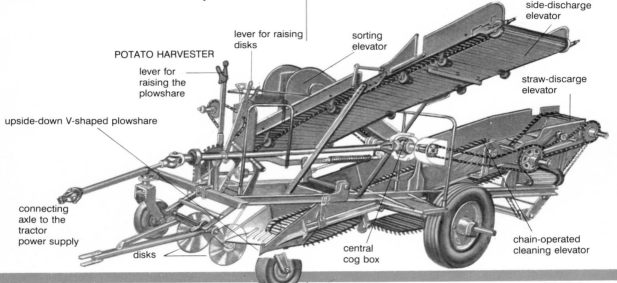

POTATO HARVESTER

lever for raising disks

lever for raising the plowshare

upside-down V-shaped plowshare

connecting axle to the tractor power supply

disks

central cog box

sorting elevator

side-discharge elevator

straw-discarge elevator

chain-operated cleaning elevator

What is a harrow?

The harrow is a machine that is used after plowing. It breaks up the thick, sticky clods of soil produced in the plowing phase. It has pointed steel teeth or rounded disks that shatter these chunks of earth, smoothing and leveling the surface. In some cases, fertilizer, pesticide or herbicide (weed killer) dispensers are attached to the harrow.

What is a spraying machine used for?

Using blades activated by a small motor, a sprayer is a machine that distributes anti-parasite powders. On the right, you see an example of one used for fruit tree protection.

Are there machines for planting seeds?

One is shown here on the right. In this machine, the distribution of the seeds is controlled by the rolling of the wheels.

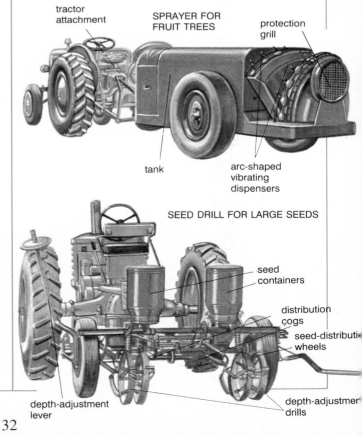

tractor attachment

SPRAYER FOR FRUIT TREES

protection grill

tank

arc-shaped vibrating dispensers

SEED DRILL FOR LARGE SEEDS

seed containers

distribution cogs

seed-distribution wheels

depth-adjustment lever

depth-adjustment drills

Are there also harvesting machines?

Yes, the harvest is done with a series of specialized machines. A combine is used to harvest most grains. This machine first cuts the plant stalks and then threshes, or separates, the grain from the straw. It leaves the straw on the ground and continues harvesting the grain which it blows into a holding tank. When the tank is full, the grain is blown into wagons or trucks to be transported to storage areas. Many modern combines are autonomous; they no longer need to be pulled by a tractor.

What is a husking machine?

It is another type of harvesting machine used for corn. It breaks off the ear of corn, removes the husk, and shucks the corn from the cob, in one quick process.

Are there also machines for spreading manure?

Yes, this, too, is an important agricultural operation since manure is a fertilizer. A machine for spreading manure is shown below. The manure, which is kept in a special box, is broken up into very fine particles by a disintegrator and is uniformly distributed by rotary blades.

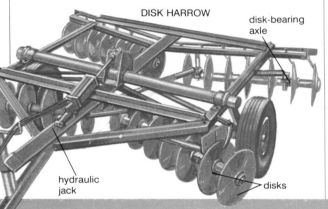

DISK HARROW

disk-bearing axle

hydraulic jack

disks

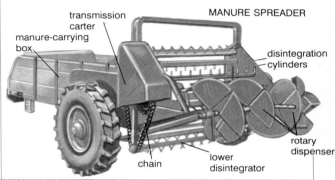

MANURE SPREADER

transmission carter

manure-carrying box

disintegration cylinders

rotary dispenser

chain

lower disintegrator

LAWN MOWERS

How did people cut grass in fields and gardens before lawn mowers were invented?
Before the middle of the last century, people used scythes and sickles. Or cows, sheep and goats were left free to graze, which kept the grass trimmed. Then mowers, consisting of a frame with a cutting bar, were introduced.

When were garden lawn movers invented?
The first domestic lawn mowers were adapted from reapers pulled by horses whose hooves were covered in burlap to prevent them from damaging the turf. Then, in 1830, Edwin Beard Budding developed the first cylinder cutter, which is the same as the kind we use today.

How does a cylinder cutter work?
The cylinder is made up of an axle with a wheel at each end which supports a series of transversal S-shaped blades. The cylinder is connected to a long U- or T-shaped handle and to a flat, fixed blade, mounted directly behind it, usually on one or two stabilizing rollers. When the operator pushes the mower, the blades rotate, trapping the grass against the fixed blade and cutting it with the same action as the two blades of a pair of closing scissors.

Is the hand-operated mower still used?
It is still used for smaller surfaces. Since the wheels that are pushed or pulled by a person move slowly, to make the blades move more quickly on the grass, the movement of the wheels is accelerated by a system of cogs.

Even though the operation of cutting the grass seems simple, there are many different types of mowers: manual or power, handstarting or electric engines, with rotary cylindrical blades or blades like those on an old paddle-wheel boat.

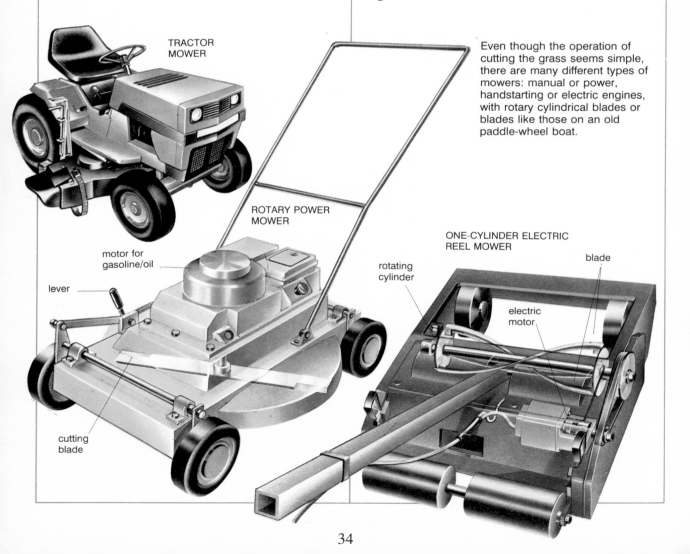

TRACTOR MOWER

ROTARY POWER MOWER

motor for gasoline/oil

lever

cutting blade

ONE-CYLINDER ELECTRIC REEL MOWER

blade

rotating cylinder

electric motor

How does the power mower work?

The most common type of power mower has a single rotary blade, mounted on a vertical axle, which cuts grass with a scythe-like motion. The blade spins rapidly beneath a metal casing that has four wheels and a handle attached. The motor is on top of the casing so that its drive shaft can turn the rotary blade below it.

Can these mowers be regulated?

Yes, they can usually be adjusted for different heights by means of a lever located either on the wheels or on the axle that controls the blade. It is often possible to adjust the speed for heavier or lighter grass as well.

Are there many different types of mowers?

Yes, there are. The most obvious difference is between hand and power mowers, but there are also mowers for different types of grass or lawn. Tractor mowers, which allow the operator to sit down and steer with greater accuracy, are especially useful for large lawns such as soccer fields, golf courses and recreation parks which require evenly-cut grass. Heavy-duty mowers exist for unusually tough grass. One type of rotary mower dispenses with wheels altogether, floating on an air cushion like a hovercraft. Lawn edgers and trimmers are used to cut grass in places where ordinary lawn mowers cannot reach, for example, around tree roots or under hedges.

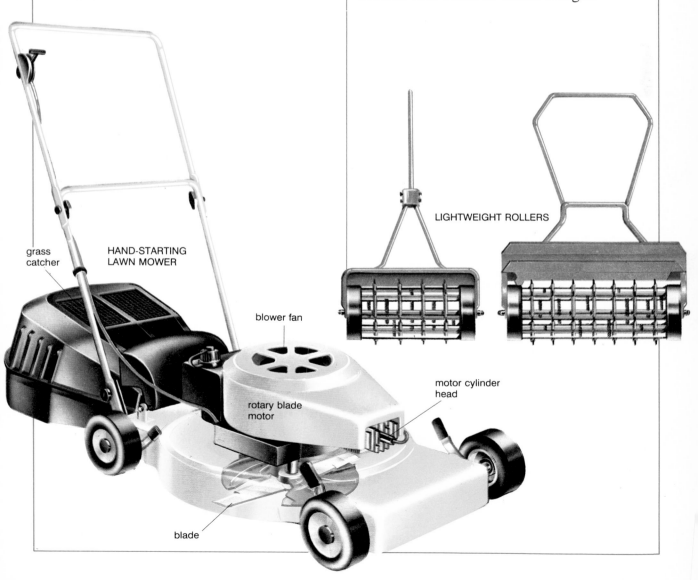

grass catcher

HAND-STARTING LAWN MOWER

blower fan

rotary blade motor

blade

LIGHTWEIGHT ROLLERS

motor cylinder head

FROM PRODUCER TO CONSUMER

A long time may pass between harvesting foods and serving them. How are they preserved in the meantime?

At one time, there were not many ways of preserving foods for a long time. Almost all produce, fruits and vegetables, as well as meat, had to be eaten in a hurry, before it spoiled. At best, these foods could be dried. Food rots because of the action of bacteria, microscopic organisms that gradually decompose organic matter. This decomposition creates bad-smelling substances which make the food not only inedible but also poisonous. But these decay-causing bacteria cannot grow at low temperatures and, therefore, the solution for preserving foods is keeping them cold.

At what temperature are foods preserved?

That somewhat depends on the type of food and the length of time you want to keep it. The "fresh" produce, like the fruits and vegetables which arrive at our tables, are kept in refrigerators at temperatures near freezing. They are transferred to the point of sales in refrigerated trucks. (The route shown by the lightly-colored arrows.) Products which have to be preserved longer are frozen at temperatures of twenty or thirty degrees below zero (the darker arrows).

What is freezing?

It is preserving products at temperatures of less than eighteen degrees below zero after rapid cooling in specialized freezers. At those temperatures, the micro-organisms that cause spoiling cannot grow, and the foods can be preserved for months. Another advantage of freezing is that once the foods are defrosted, they have an appearance, taste and nutritional value which is similar to that of fresh foods. Frozen products, however, once taken out of the freezer, cannot be exposed to temperatures of more than eighteen degrees below zero for very long.

food products

refrigerated trucks

SUPERMARKET

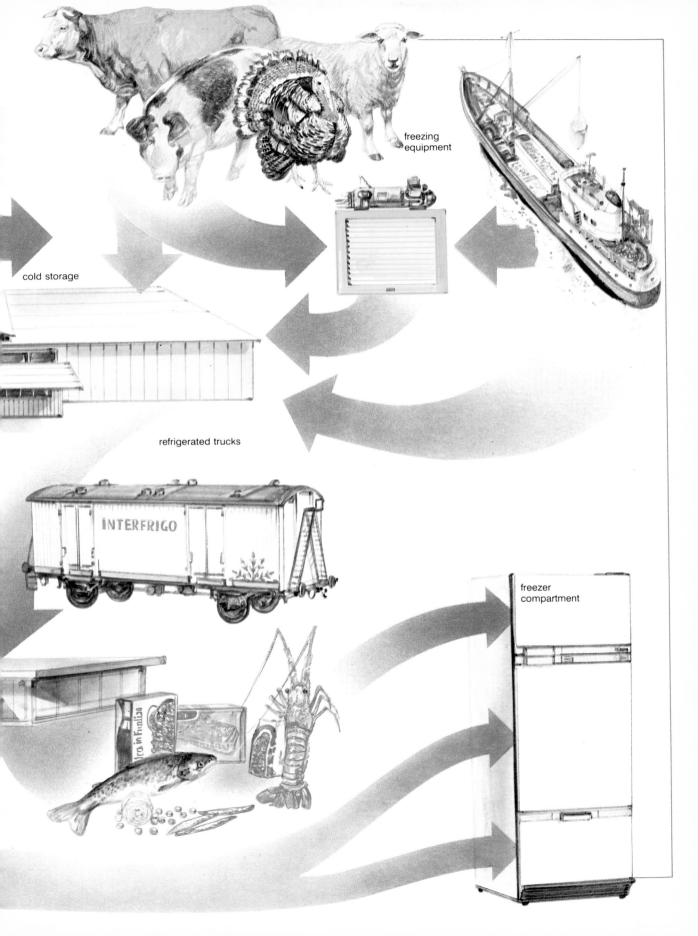

freezing
equipment

cold storage

refrigerated trucks

INTERFRIGO

freezer
compartment

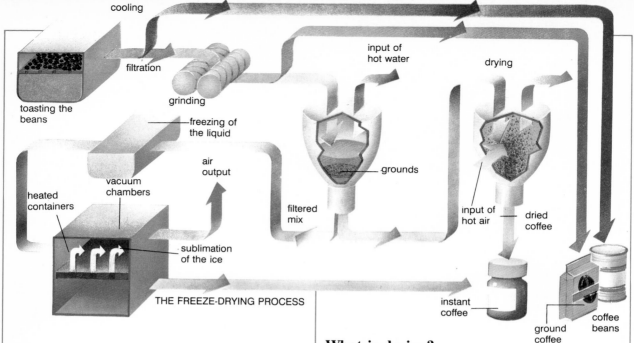

cooling

filtration

toasting the beans

grinding

freezing of the liquid

input of hot water

drying

heated containers

vacuum chambers

air output

grounds

sublimation of the ice

filtered mix

input of hot air

dried coffee

THE FREEZE-DRYING PROCESS

instant coffee

ground coffee

coffee beans

FOOD PRESERVATION

Is freezing the only way to preserve foods?
Although it is common, being very natural and requiring few modifications to the food, freezing is not the only technique used. Others include salting and drying, techniques that were practiced centuries ago, and canning and freeze-drying, methods that are very new.

Which foods are canned?
Many foods are canned, including beef, tuna, sardines and vegetables, like tomatoes. (The photograph on the opposite page shows a detail of the tomato canning industry.) In the case of meat (see the chart at the top of the opposite page), after it has been de-boned, cut, cooked and sorted, it is hermetically closed in metal cans so that it is not in contact with the air. Then the cans are sterilized at high temperatures (over one hundred degrees) to eliminate bacteria and microorganisms.

For how long is canned food preserved?
Food in cans can last a long time, even for years.

What is drying?
It is a technique used for eliminating a liquid (usually water) from a solid, by evaporation. Drying has been traditionally used for food preservation. Without water, the organisms that cause spoiling develop very slowly or sometimes not at all.

Which foods are usually dried?
We still preserve cereals, fruit (usually figs and prunes) and mushrooms in this way.

How are foods dried?
Nowadays, they are dried in special devices called dryers, usually by means of a current of hot air which is passed through the product or by passing the product itself in front of a heat source.

What is freeze-drying?
Freeze-drying is a preservation technique which works by taking the water out of foods (that is, it dehydrates them), without altering their basic characteristics. When you want to use freeze-dried foods, you put them in water and they take on the same appearance and properties that they had at the beginning.

Which products are freeze-dried?
Examples are meat, fish, milk, coffee and fruit juices.

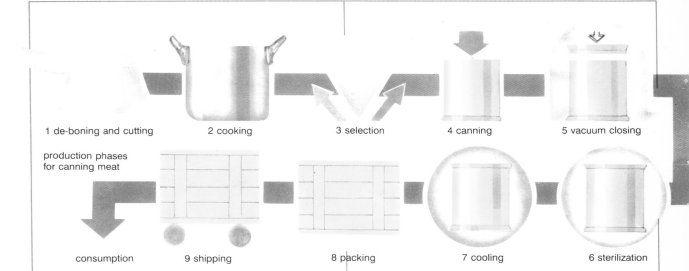

1 de-boning and cutting 2 cooking 3 selection 4 canning 5 vacuum closing

production phases for canning meat

consumption 9 shipping 8 packing 7 cooling 6 sterilization

How does the freeze-drying process work?

The foods to be freeze-dried are put into a special apparatus called an autoclave, placed on a series of plates which can be heated, by a warming system, or cooled, by a refrigeration system. After the autoclave has been closed, a pump is used to create a vacuum in it. At the same time, the foods to be freeze-dried undergo very quick freezing, by means of the cooling plates. In this way, the water in them solidifies. At this point, the plates are slightly heated. Since the water is in a vacuum, it changes directly from a solid state to a gaseous state. The gas is then collected in a condenser. The foods, almost completely dehydrated and unrecognizable, are then packaged, ready to be sent to the consumer. Freeze-dried products can be preserved for a very long time.

Are there foods that are preserved in more than one way?

Almost all food products are preserved in more than one way in order to satisfy different needs. Vegetables, for example, are preserved in the refrigerator for "fresh" use, but they can also be frozen or canned. Coffee (see the chart at the top of the opposite page), after being toasted, can be preserved as beans or can be ground or freeze-dried (instant coffee).

IRRIGATION

What is irrigation?
It is the process of watering cultivated land in order to help the plants grow, to protect them from dryness, and to improve both the quantity and the quality of the produce.

Is it done by natural methods?
In many kinds of terrain, irrigation can be done naturally, by diverting some of the water of rivers or streams into canals. But some form of artificial irrigation is needed in many areas when, for example, there are no convenient sources of water or there is an unfavorable distribution of water in the different seasons. It may also be necessary when the terrain itself is not suitable or is dry.

How is subsurface irrigation done?
Subsurface irrigation brings the water to the plants directly from the soil. For this system to work, the land must be level, extremely permeable on the surface and impermeable in the lower levels, so that the water can be absorbed slowly and not dispersed into the soil.

How does surface irrigation work?
This is the most common method of irrigation. It can be used for different types of terrain. The land may be flooded (with the surfaces to be irrigated divided into fields with a single water source or into two fields which slope away from each other). There may be furrows or canals or, if the land slopes, there may be a main canal with a natural flow, for spontaneous infiltration. These three methods are shown in the drawings on the left.

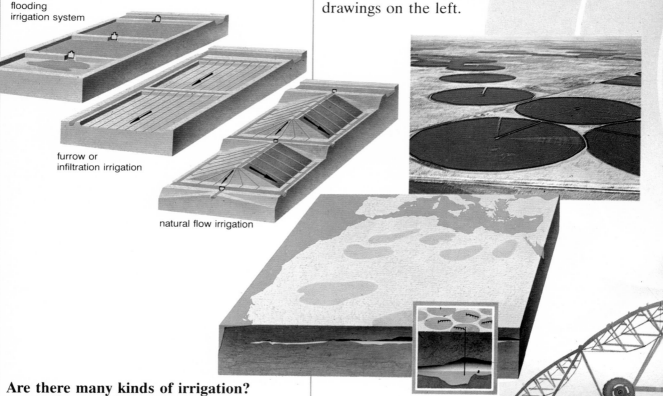

flooding
irrigation system

furrow or
infiltration irrigation

natural flow irrigation

Are there many kinds of irrigation?
There are three main methods of irrigation currently in use: underground or subsurface, surface and sprinkling. The choice of one method over another depends upon the characteristics of the land and the extent and type of cultivation.

How does sprinkling work?
Sprinklers of different sizes, depending on the extent and type of cultivation, are used. They are connected to pumps, which distribute the water to the plants.

What is drip irrigation?

It is another, recently-developed, system of irrigation, which is very ingenious and very expensive. Because of its high cost, it is not used much. The water is carried to the roots of the plants by thin plastic tubes. The plant receives a regular dose of water from a spout. This technique is most often used for fruit cultivation.

Are agricultural techniques changing?

Although agriculture is already quite mechanized, more changes can be predicted because the food needs of the world continue to grow. There are already those who think about using robots, with no human intervention, for work in the fields.

Can plants grow in the desert?

Yes. Experimental cultivation has been tried in the desert, using special irrigation techniques, like the mobile arm (shown below). A mobile arm, which can be miles long, is used for large areas. It is on wheels and has many nozzles for water distribution. With techniques like this, profitable cultivation has been done in the Sahara.

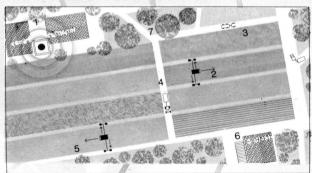

AGRICULTURAL MACHINES OF THE FUTURE

1 central control unit
2 working robot
3 inactive robot
4 inactive robot
5 working robot
6 storage and spare part centers
7 irrigated areas

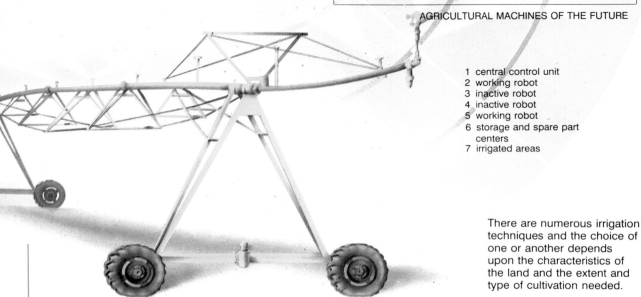

There are numerous irrigation techniques and the choice of one or another depends upon the characteristics of the land and the extent and type of cultivation needed.

THE FUTURE OF AGRICULTURE

Why do we try to intervene at the cellular level in plants?

Farmers have long tried to cross different species to get varieties that are more resistant to particular climatic conditions or that can produce greater quantities, larger sizes or other particular characteristics. Cross-breeding has the effect of producing varieties with intermediary characteristics. The results depend on what happens to the genetic characteristics (the microscopic structures on which the visible characteristics of the organism depend) in the cells of the plants.

Why directly manipulate seed cells?

Because with hybridization and cross-breeding, you never quite know what will happen. The result depends upon the chance recombination of the genetic material. By directly manipulating the cells of the seeds, as some genetic engineers do, you get the results you would like.

Is this difficult to do?

Yes, it is. Many genetic modifications can make a plant competitive. For this reason, plants have many properties and, therefore, many genes to which interventions must be made to get the desired change. Moreover, the equilibrium of plants is complex, and the result of an experiment is certain only after the new plant form has matured.

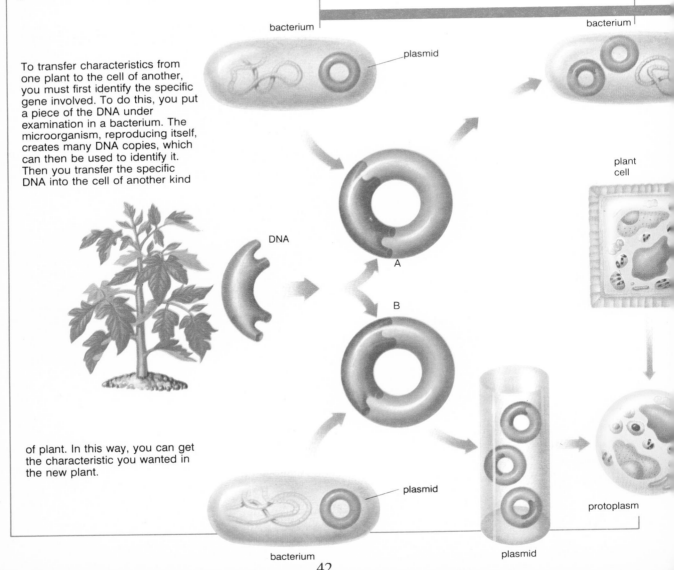

To transfer characteristics from one plant to the cell of another, you must first identify the specific gene involved. To do this, you put a piece of the DNA under examination in a bacterium. The microorganism, reproducing itself, creates many DNA copies, which can then be used to identify it. Then you transfer the specific DNA into the cell of another kind of plant. In this way, you can get the characteristic you wanted in the new plant.

bacterium

plasmid

bacterium

plant cell

DNA

A

B

plasmid

bacterium

plasmid

protoplasm

Are there other research areas in which work is being done to improve agriculture?
Yes, for example, improvements can come from studying the best conditions for the cultivation of particular species to get more and better produce. (The photograph on the right, for instance, shows an experimental greenhouse.) More efficient irrigation techniques could also give good results and could allow for the cultivation of currently unproductive land. We must, however, be careful because there are also agricultural techniques which give good results in the short run but which, in the long run, can deplete the soil to a point that it becomes unfit for cultivation. Decades may pass before these areas can be planted again.

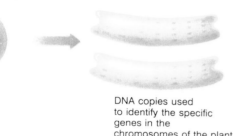

DNA copies used to identify the specific genes in the chromosomes of the plant

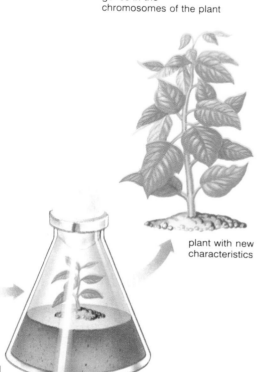

plant with new characteristics

plant cell in culture

Is genetic engineering applied only to agriculture?
Genetic engineering research deals with organisms in general. It could be applied not only to plants but also to animals (with regard to breeding) and to humans (where it may have important medical implications).

43

DID YOU KNOW...?

What is agronomy?

Agronomy is the discipline that studies the application of electronics and automation to agriculture. Using a computer to control a greenhouse or to determine the best conditions for planting and harvest comes under the field of agronomy. As in all other areas of life, the computer has recently become more and more important in agriculture, both as an aid for administration and management of farm businesses and as an instrument for controlling automatic processes.

What is the Green Revolution?

The Green Revolution is a name that refers to the application of modern technology to agriculture, which has occurred principally in the second half of the twentieth century. The Green Revolution has allowed for a significant increase in the production per unit of cultivated land.

What is the horsepower of a tractor engine?

Only thirty years ago, thirty horsepower engines were considered real marvels, but today most modern tractors reach a horsepower of more than three hundred.

Do all tractors have wheels?

No, they don't. There are also caterpillar tractors for demanding and difficult work like clearing the land or working soil that is extremely bumpy or soft. The tracks, or caterpillars, of these tractors are similar to those of military tanks.

Are tractors used in sectors other than agricultural?

Yes, there are tractors for industrial or building use which are based on the same principles as four-wheel or caterpillar farming tractors. They are, however, heavier, stronger and faster. They are chiefly used to pull and operate construction and excavation equipment, such as back hoes, scrapers and loading machines.

How is a combine made?

The combine does both the harvesting and threshing (separating the wheat from the chaff) which, at one time, involved three different operations and the labor of many people. The combine is powered by a simple hand-starting engine which is located in the box behind the driver. The drive, which controls the wheels and all the mechanisms of the machine, is activated by a vertical pulley. The engine also controls a hydraulic pump for the power-operated wheel, a kind of large conveyor rake which directs the harvested grain toward the threshing complex awner, or beater. It also controls the cutting platform and the mechanism for changing the speed. Two large grain separators at the front direct the harvested grain in preset quantities to avoid overloading the threshing complex.

Do milking machines exist?

Yes, they do. In modern barns, even milking can be done mechanically. A carousel milking machine is an especially complicated device. It consists of a large, slow-moving platform, powered by an electric motor, with places for several cows. As the cows approach the platform, they are placed at their own feeders, and a milking machine is attached to each cow. The milk produced is collected in glass containers.

How is the milking done?

Tubes with rubber membranes are put on the udder of the cow. The membrane is placed on the teat, and a device alternatively squeezes the tube and releases air into it, thus closing and opening the membrane. This movement imitates the sucking of a calf. But the milk, in this case, is sucked into a tube by which it is transported to the containers.

How much milk does a cow produce?

On the average, a European cow produces about 2,850 quarts (2,700 l.) of milk a year, though some animals can produce three times as much.

DOMESTIC
TECHNOLOGY

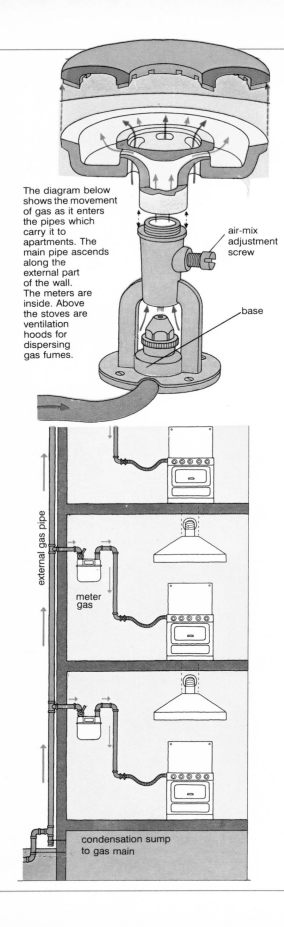

The diagram below shows the movement of gas as it enters the pipes which carry it to apartments. The main pipe ascends along the external part of the wall. The meters are inside. Above the stoves are ventilation hoods for dispersing gas fumes.

air-mix adjustment screw

base

external gas pipe

meter gas

condensation sump to gas main

STOVES AND OVENS

How do gas burners work?
The flammable gas circulates in ring-shaped burners which are controlled by valves. When the gas is lit, the valves control the height of the flame.

Where does the gas come from?
It may come from a gas cylinder or, as is usually the case in cities, from a public distribution network. Private homes are connected to this network by special pipes.

Why are ovens useful?
Both gas and electric burners heat only from the bottom. In a closed oven, however, the heat can reach the food from all directions at the same time.

How does a microwave oven work?
Gas or electric ovens cook food by producing heat from gas combustion or from current passing through the resistance. Microwave ovens, instead, cook food by means of radiation. This radiation is generated by a tube called a magnetron. It makes the food molecules vibrate, creating heat by the internal friction that results from the collision of these molecules.

Are microwave ovens dangerous?

Microwaves can be dangerous if they reach the human body. Manufacturers, as a rule, guarantee that the oven walls completely prevent the radiation from escaping and damaging the person using the microwave.

In a microwave oven, electromagnetic waves passing through the food continuously invert the position of the molecules. This molecular disturbance causes the food to heat up.

Food molecules have electric charges turned in every possible direction.

A microwave impulse causes the molecules to align themselves parallel to their magnetic field.

The next impulse makes them line up in the opposite direction.

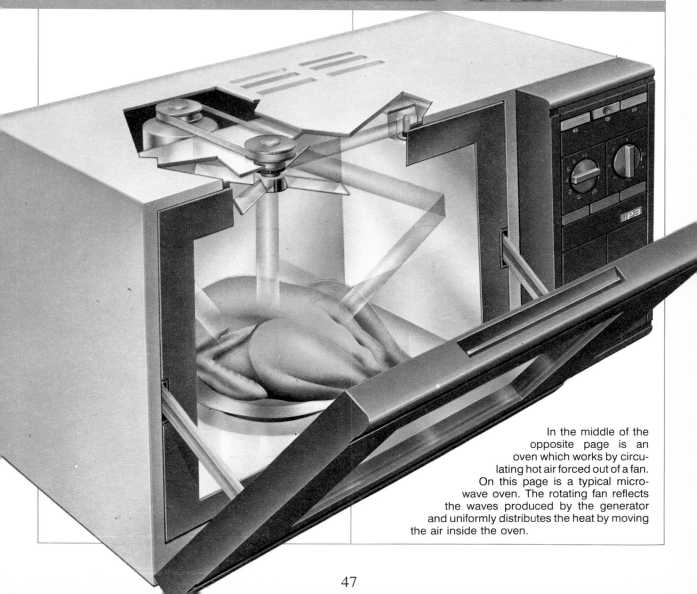

In the middle of the opposite page is an oven which works by circulating hot air forced out of a fan. On this page is a typical microwave oven. The rotating fan reflects the waves produced by the generator and uniformly distributes the heat by moving the air inside the oven.

WASHING MACHINES AND DISH-WASHERS

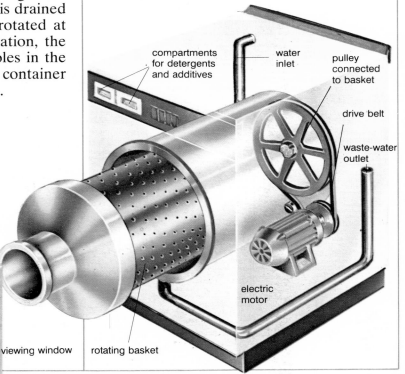

tank

basket

anti-vibration frame

How does a washing machine clean clothes?

The basic method is to clean clothes by forcing water and detergent through them. The rapid movement of the machine substitutes for the arms used in hand washing. An agitator in a closed space is one way to obtain this movement. Another is to rotate the wash in a cylinder equipped with fins.

What does "centrifugation" mean?

It is the procedure used at the end of the washing and rinsing cycles to get rid of most of the water in the clothes, making them almost dry. First the rinse water is drained and then the clothes basket is rotated at high velocity. Forced by the rotation, the water exits through the small holes in the basket and goes into an external container from which it is then discharged.

Does a dishwasher work like a washing machine?

Not really because the dishes, unlike the clothes, must not move or they will break. The washing, however, follows the same principle, running water and detergent over the dishes with sufficient force to clean them. In this case, a pump is used to input the water.

compartments for detergents and additives

water inlet

pulley connected to basket

drive belt

waste-water outlet

electric motor

viewing window

rotating basket

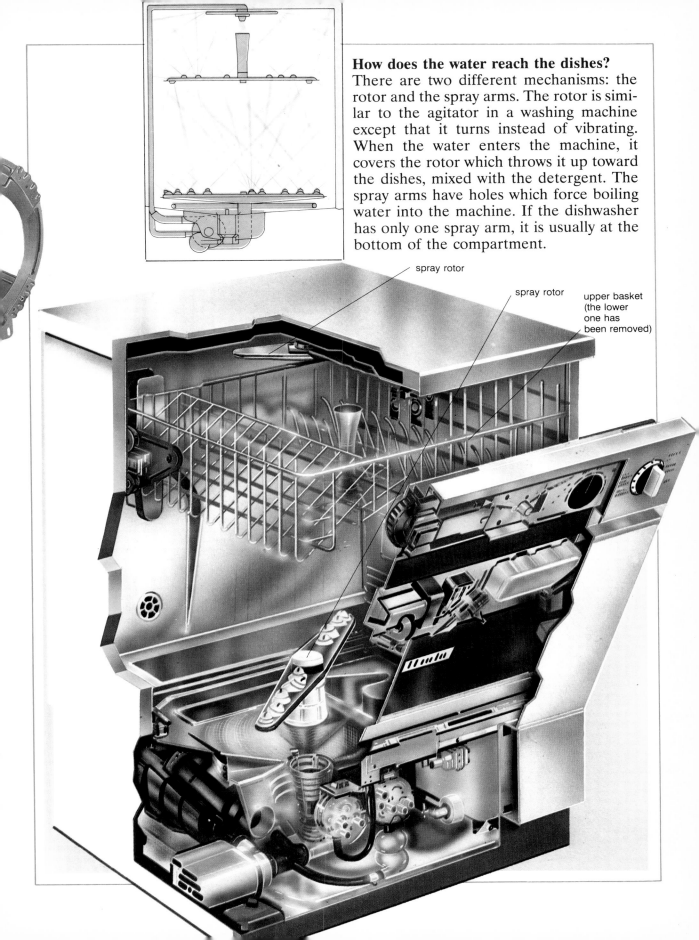

How does the water reach the dishes?

There are two different mechanisms: the rotor and the spray arms. The rotor is similar to the agitator in a washing machine except that it turns instead of vibrating. When the water enters the machine, it covers the rotor which throws it up toward the dishes, mixed with the detergent. The spray arms have holes which force boiling water into the machine. If the dishwasher has only one spray arm, it is usually at the bottom of the compartment.

spray rotor

spray rotor

upper basket (the lower one has been removed)

VACUUM CLEANERS

How do vacuum cleaners work?
The basic principle is the use of a motorized fan which sucks in the air outside of the machine. The particles of dirt and dust that come in contact with the vacuum cleaner are carried by the air current into the machine.

Where is the dirt deposited?
It is usually collected in a filter bag just beyond the entry point.

Does the air also stay in the bag?
No, the bag is very porous so that the air can escape. It then flows out of the machine through an opening in the back of the vacuum cleaner where the motor and the other components are also located.

Doesn't dirt escape with the air?
The bag is porous but not to that extent. The holes are the smallest possible for the type of material used and for the strength of the aspiration. If this were not true, the finer dust, and with it, the bacteria, would again escape into the room where they might be inhaled.

What are the bags made of?
Bags may be made of cloth or of reinforced paper. Those made of cloth are emptied and reused; those made of paper, instead, are thrown out and replaced when they become full.

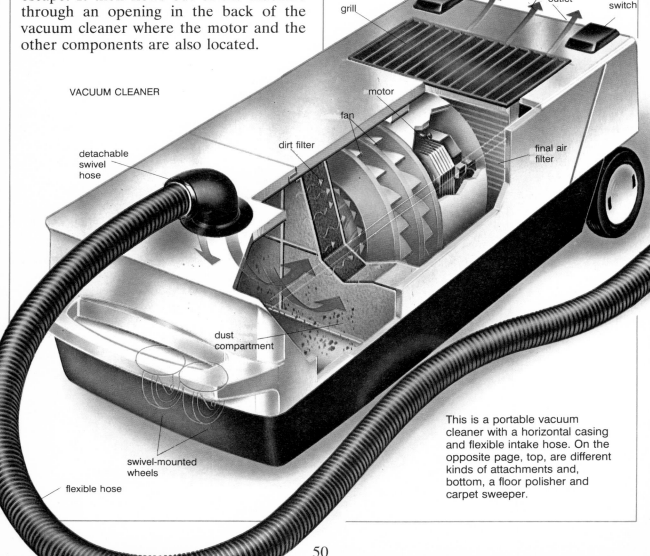

VACUUM CLEANER

grill

air outlet

switch

motor

fan

final air filter

dirt filter

detachable swivel hose

dust compartment

swivel-mounted wheels

flexible hose

This is a portable vacuum cleaner with a horizontal casing and flexible intake hose. On the opposite page, top, are different kinds of attachments and, bottom, a floor polisher and carpet sweeper.

50

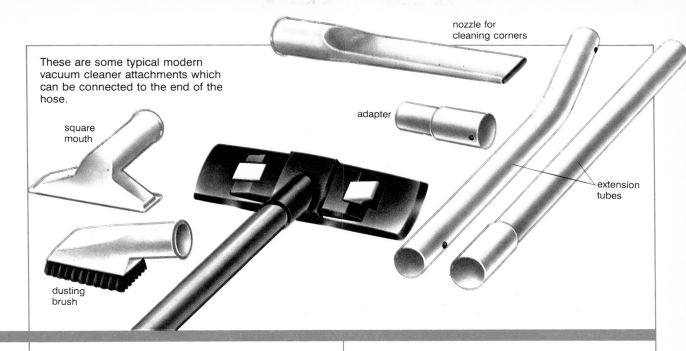

These are some typical modern vacuum cleaner attachments which can be connected to the end of the hose.

nozzle for cleaning corners

square mouth

adapter

extension tubes

dusting brush

Does anything happen as the bag fills up?

The sucking power of the vacuum cleaner progressively diminishes as the bag fills up. When the bag is almost full, the decrease may be considerable. At this point, it's a good idea to change the bag to avoid wasting electrical energy. In newer models, a warning light goes on automatically when the bag has to be changed.

Does the vacuum cleaner only clean surface areas?

That is, of course, its main function. But modern vacuum cleaners are usually equipped with numerous attachments which are connected to the hose. These can be used for more specialized jobs, such as cleaning in cracks.

Are there many attachments that can be used?

Yes, although in the illustration above, you see only a few of them. They may be simple extension tubes for reaching places farther from the machine or brushes for removing dust which is later picked up. Narrow nozzles increase the vacuum's power and can be used to pick up heavier dirt particles. Attachments for cleaning non-carpeted floors tend to be relatively narrow and to have fine bristles. Attachments with long, soft bristles are used for dusting furniture.

Does a floor polisher work like a vacuum cleaner?

Floor polishers usually have an aspiration unit to make sure that the dust does not stick to the polishing mixture.

FLOOR POLISHER

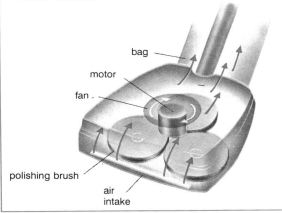

bag

motor

fan

polishing brush

air intake

VACUUM CLEANER

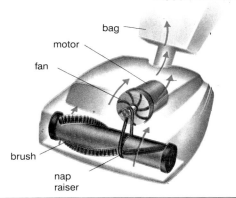

bag

motor

fan

brush

nap raiser

HAIR DRYERS AND ELECTRIC RAZORS

How does a hair dryer work?
It produces hot air and directs it toward the hair.

What element heats the air?
The resistance, located in the air duct that sucks air in from outside, warms up as current passes through it.

How is the air moved?
A small metal or plastic fan brings the air toward the hair dryer as it turns, activated by an electric motor with a high number of revolutions per minute.

How does an electric razor work?
An electric mechanism sets in motion one or more ridged blades. Above the blade there is a metal head which is also ridged. The facial hair penetrates this head and is cut by the moving blade.

head with perforated blade

blade

vibrator

electric coils

How does the blade move?
The alternating current supplied to the razor polarizes an electromagnet which alternatively attracts and repels the T-shaped bar holding the blade. Since the current alternates, it changes sign several times a second, producing a very quick movement of the blade.

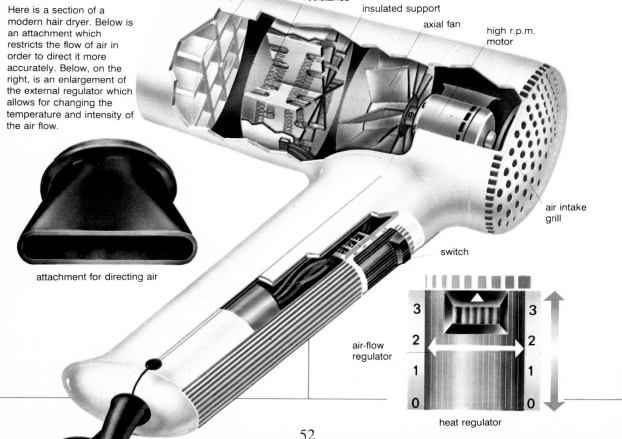

Here is a section of a modern hair dryer. Below is an attachment which restricts the flow of air in order to direct it more accurately. Below, on the right, is an enlargement of the external regulator which allows for changing the temperature and intensity of the air flow.

grill

resistance

insulated support

axial fan

high r.p.m. motor

air intake grill

attachment for directing air

switch

air-flow regulator

heat regulator

3 3
2 2
1 1
0 0

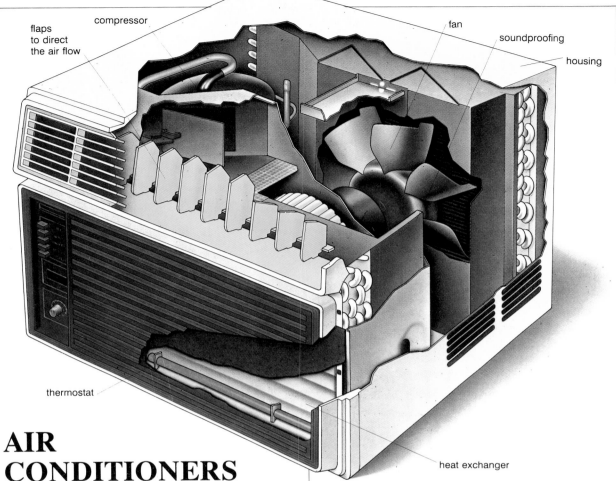

flaps to direct the air flow

compressor

fan

soundproofing

housing

thermostat

heat exchanger

AIR CONDITIONERS AND ELECTRIC HEATERS

What does an air conditioner do?
It keeps the air temperature within given limits, controls the humidity, and removes dust, pollen and other particles from the air.

How is the air cooled?
One of a number of cold-generating fluids derived from Freon 19 is run through a coil, and the air to be cooled is brought into contact with this coil by a fan. The fluid absorbs the heat and, since it has a low boiling point, becomes gaseous. The cooled air returns to the room while the refrigerant is sent to a compressor and a condenser from which its heat is discharged into the cooler air outside. This cycle then repeats itself.

How does an electric heater work?
A resistance, which overheats as current passes through it, heats the air. The conventional type of heater also has a fan which circulates the hot air to heat up the area as quickly as possible.

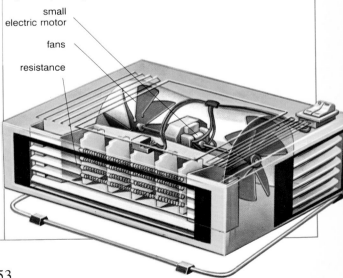

small electric motor

fans

resistance

FOUNTAIN PENS

What is a fountain pen?
A fountain pen is a writing instrument which contains its own ink supply. At one time, even as late as the twentieth century, pens had to be continually refilled from external inkwells. Today, when we use the expression "fountain pen," we usually mean a pen which ends in a point, or nib.

What are the parts of a fountain pen?
A fountain pen has many important parts: a nib for writing; a reservoir which holds the ink; a tube which carries the ink to the nib; a feeder which regulates the flow of the ink; the housing, the small cylinder that contains the pen; and a cover for protecting the nib.

What is a refillable reservoir?
It is a small, flexible rubber bag which contains the reserve ink.

How is it loaded?
A small lever, or another similar device, pushes the air out of the reservoir. Then, when you put the pen in ink and release the lever, the ink is sucked up into the reservoir. Instead of a lever, the device may be a screw which pushes a piston. In other models, the pressure on the reservoir is created manually. In any case, the point is to create a vacuum in the reservoir so that the pen can be filled with ink by using the process of aspiration.

What is happening when the reserve empties as you write?
The connection between the nib and the reservoir is a network of tiny passages that carry air to the empty space at the top of the reservoir. As you write, the air replaces the ink that flows to the nib.

What is the feed?
The feed, found behind the nib, is a piece of hard rubber or plastic with comblike slots in it. This component regulates the ink flow by storing any excess ink that may travel down from the reservoir.

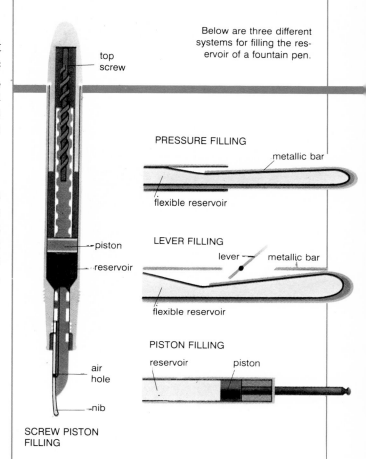

Below are three different systems for filling the reservoir of a fountain pen.

top screw

piston

reservoir

air hole

nib

SCREW PISTON FILLING

PRESSURE FILLING

metallic bar

flexible reservoir

LEVER FILLING

lever   metallic bar

flexible reservoir

PISTON FILLING

reservoir   piston

What are pens with ink cartridges?
They are special fountain pens that have their ink reserve in a small plastic container (the cartridge) which is placed inside the pen when the previous cartridge is empty. The refill cartridge is replaced, making enough ink available for a considerable amount of writing. This type of refilling does not use any mechanical devices and, therefore, the pen lasts longer. Moreover, this pen has the advantage of being convenient and simple. For all of these reasons, cartridge pens are the most popular type today.

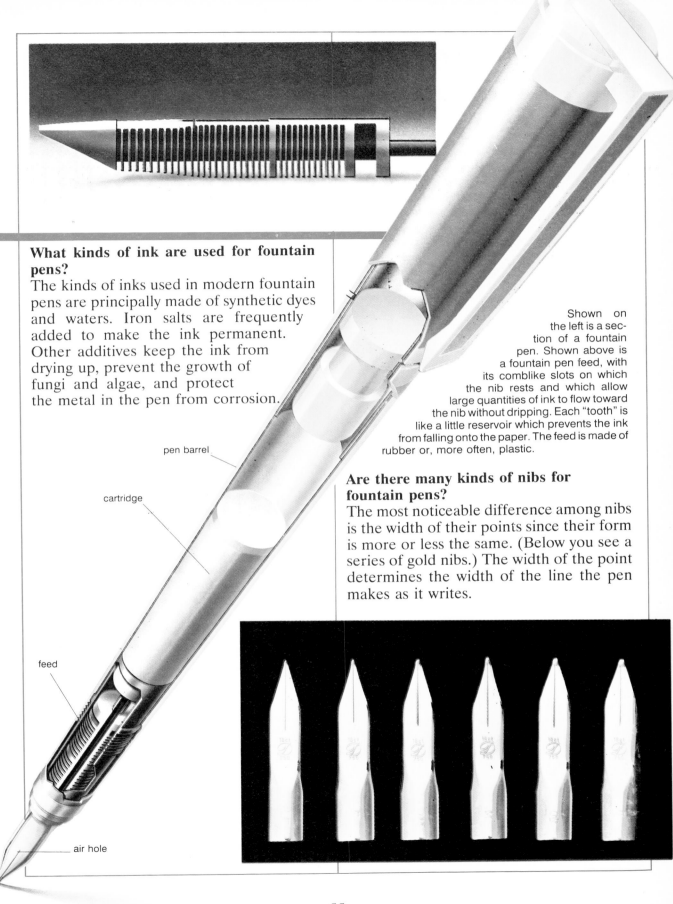

What kinds of ink are used for fountain pens?

The kinds of inks used in modern fountain pens are principally made of synthetic dyes and waters. Iron salts are frequently added to make the ink permanent. Other additives keep the ink from drying up, prevent the growth of fungi and algae, and protect the metal in the pen from corrosion.

Shown on the left is a section of a fountain pen. Shown above is a fountain pen feed, with its comblike slots on which the nib rests and which allow large quantities of ink to flow toward the nib without dripping. Each "tooth" is like a little reservoir which prevents the ink from falling onto the paper. The feed is made of rubber or, more often, plastic.

Are there many kinds of nibs for fountain pens?

The most noticeable difference among nibs is the width of their points since their form is more or less the same. (Below you see a series of gold nibs.) The width of the point determines the width of the line the pen makes as it writes.

pen barrel

cartridge

feed

air hole

BALLPOINT PENS

How do ballpoint pens work?
The operation of ballpoint pens is based upon a rolling iron sphere or ball, which distributes the ink onto the paper from a reservoir in the pen.

What kind of ink do they use?
Ballpoint pens are designed to use an ink that dries quickly and does not drip or stain. This ink begins to flow at the moment you begin writing. Enough ink can be contained in the pen to write for a long time without needing a replacement. The ink is usually thick and gelatinous.

How is a ballpoint pen made?
A ballpoint pen has three parts: the shell or external part, the reservoir and the point. The plastic or metal shell contains the reservoir which may also be made of metal or plastic filled with ink. Sometimes the reservoir is sealed with a grease plug, which follows the movement of the ink toward the point so that it doesn't leak out of the top.

How is the point made?
The point is made of a little iron ball from which the name for this type of pen was taken. The ball is contained in a shell but at least half of it must be out in order to write.

How does the ball work?
Since the ball turns freely, its upper part, inked from the reservoir, can swivel to spread ink on the paper. Good ballpoint pens have a precisely ground ball that is spun into the housing. Four to six shallow grooves in the housing assure that the ink is distributed evenly on the ball.

How does the ink come out of the reservoir?
The ink descends from the reservoir to the tip by gravity and is then drawn onto the ball by natural capillary action, a property of liquids in containers with small diameters. This is why your pen will not write when you hold it to a wall, for example. If the pen is not more or less vertical, the ink cannot reach the ball.

A typical ballpoint pen, with its point, a steel ball, is shown here. The ink is in a reservoir, which may be made of metal or plastic and can be replaced when the ink runs out.

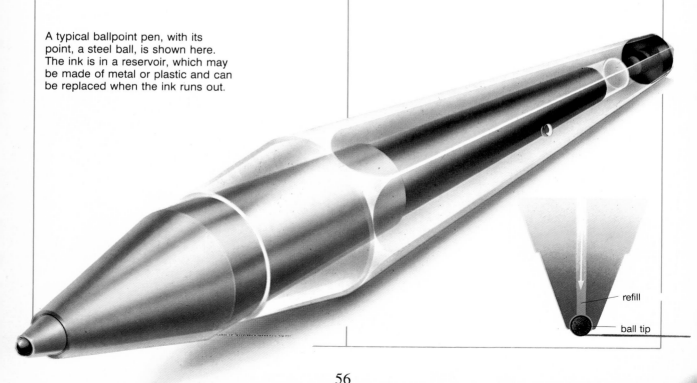

refill

ball tip

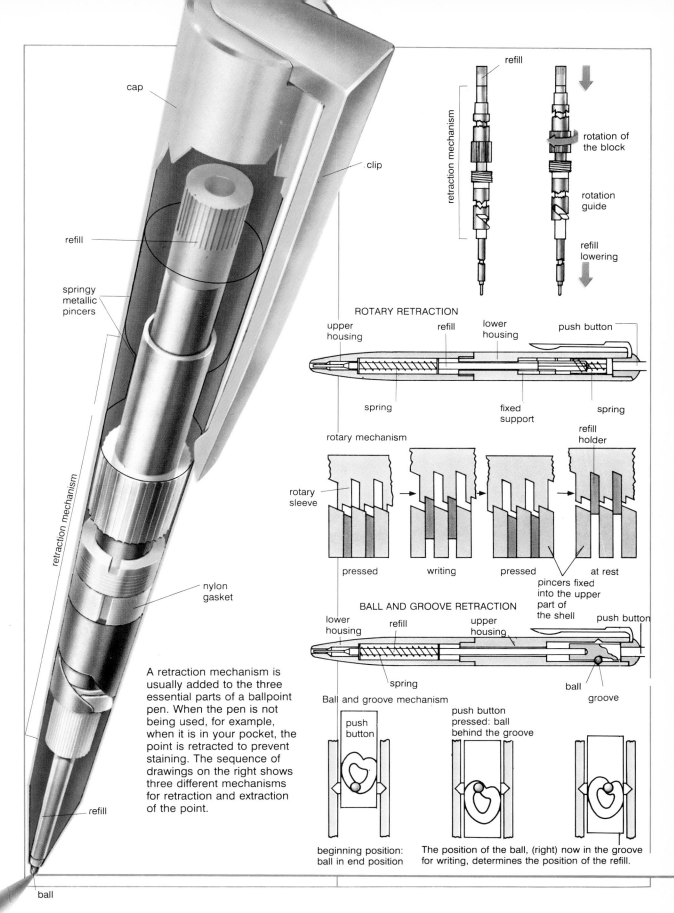

cap

clip

refill

refill

springy
metallic
pincers

retraction mechanism

nylon
gasket

refill

ball

retraction mechanism

rotation of
the block

rotation
guide

refill
lowering

ROTARY RETRACTION

upper
housing

refill

lower
housing

push button

spring

fixed
support

spring

rotary mechanism

refill
holder

rotary
sleeve

pincers fixed
into the upper
part of
the shell

pressed   writing   pressed   at rest

BALL AND GROOVE RETRACTION

lower
housing

refill

upper
housing

push button

spring

ball

groove

Ball and groove mechanism

push
button

push button
pressed: ball
behind the groove

beginning position:
ball in end position

The position of the ball, (right) now in the groove
for writing, determines the position of the refill.

A retraction mechanism is
usually added to the three
essential parts of a ballpoint
pen. When the pen is not
being used, for example,
when it is in your pocket, the
point is retracted to prevent
staining. The sequence of
drawings on the right shows
three different mechanisms
for retraction and extraction
of the point.

LIGHTERS

What is a lighter?
It is simply a miniature ignition device.

How is it made?
Every lighter contains fuel, a regulator which dispenses a suitable quantity of fuel, and a sparking system which ignites the fuel.

How do wick lighters work?
The first wick lighter was introduced in France around 1920. In these lighters, the sparks set fire to a small cord soaked in saltpeter. The cord burned slowly but steadily enough to light cigarettes. Subsequent models, instead, used wicks similar to those in oil lamps. The wick was soaked in benzene, a flammable liquid derived from petroleum, and it burned slowly. The flame was extinguished by lowering a cap over it which remained there as long as the lighter was not being used. These first lighters absolutely required the use of both hands. Around 1920, a lighter that could be operated with one hand was developed in Germany for veterans who had lost the use of an arm.

When did the first gas lighters appear?
The first lighters to use a gas such as butane as a fuel were produced after the second world war.

How do they work?
Butane, derived from natural gas, will burn in the air if ignited. In modern lighters, it is pressurized in a tiny gas tank inside the lighter. When the lighter is used, the gas shoots out of a small nozzle at the top where it is ignited by the sparking device. The nozzle controls the rate of the gas flow which, in turn, affects the size of the flame. The screw on the lighter is used to regulate the nozzle.

How does piezoelectric sparking work?
When certain crystals or ceramics receive a strong impact from the action of a small lever or button, a high-voltage electric charge is generated. The crystals produce a quick spark which is, however, enough to ignite the butane.

How are sparks created?
There are currently three methods. The oldest is the one in which a wheel strikes a flint of pressed antimony powder or a steel-cerium alloy to produce the sparks. The flint wears out with use but can easily be replaced. The other two sparking systems use piezoelectricity or batteries.

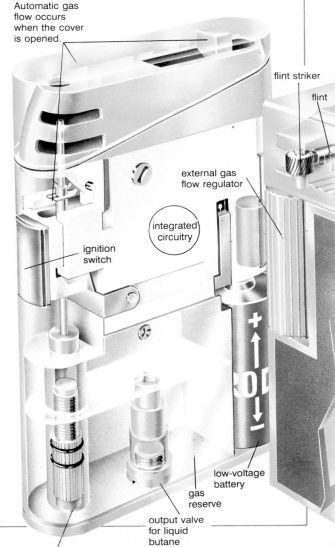

Automatic gas flow occurs when the cover is opened.

flint striker

flint

external gas flow regulator

integrated circuitry

ignition switch

low-voltage battery

gas reserve

output valve for liquid butane

flame-height regulator

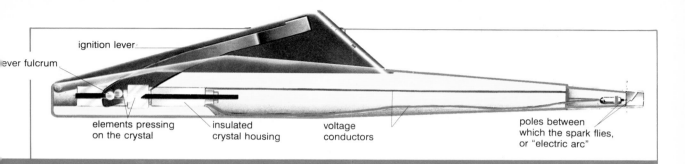

ignition lever

lever fulcrum

elements pressing
on the crystal

insulated
crystal housing

voltage
conductors

poles between
which the spark flies,
or "electric arc"

How do you refill a gas lighter?

Usually there is a valve located at the bottom of the lighter. Using a special nozzle, you attach a refill to the valve. The procedure is similar to putting air in a flat tire by using a cylinder of pressurized air.

How are battery-powered lighters ignited?

Those with a manganese battery have a miniature transformer inside. When the switch is activated, the battery raises the voltage enough to create a spark to ignite the fuel. Other models use a silver battery and integrated circuitry to create a high-voltage spark.

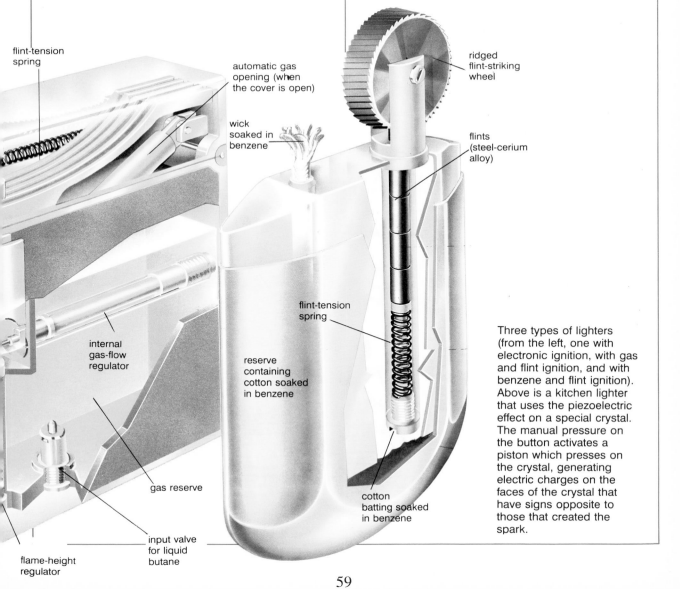

flint-tension
spring

automatic gas
opening (when
the cover is open)

wick
soaked in
benzene

ridged
flint-striking
wheel

flints
(steel-cerium
alloy)

internal
gas-flow
regulator

flint-tension
spring

reserve
containing
cotton soaked
in benzene

gas reserve

cotton
batting soaked
in benzene

input valve
for liquid
butane

flame-height
regulator

Three types of lighters (from the left, one with electronic ignition, with gas and flint ignition, and with benzene and flint ignition). Above is a kitchen lighter that uses the piezoelectric effect on a special crystal. The manual pressure on the button activates a piston which presses on the crystal, generating electric charges on the faces of the crystal that have signs opposite to those that created the spark.

LOCKS

What is the most common type of lock?
The lock found on most apartment doors is called a pin tumbler, patented by Linus Yale in 1848. It consists of a rotating cylinder and a flat key with a serrated edge.

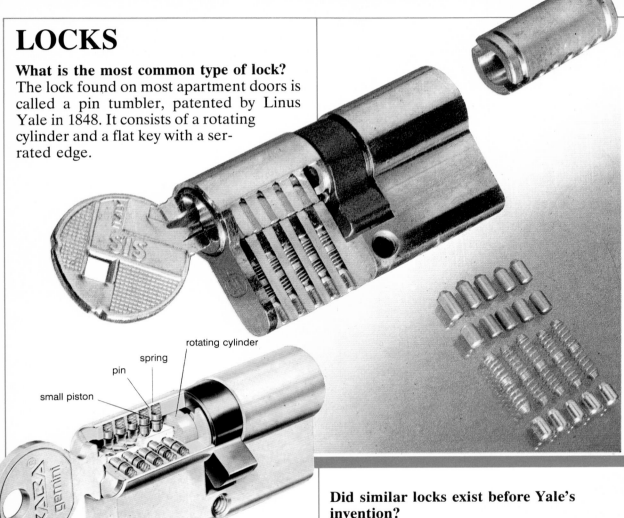

small piston

pin

spring

rotating cylinder

How does it work?
To free the cylinder so that it can rotate, the key must lift a series, usually five, of pins.

How are these pins made?
These metal pins, which are forced down into the cylinder by springs, are divided in two parts. The cylinder rotates only when the edge between the upper section and the lower section of each pin is aligned with the top of the cylinder.

How is it possible to make so many different kinds of keys?
Because the pins can be of different lengths, there are millions of possible combinations.

Did similar locks exist before Yale's invention?
Yes, the earliest lock known, made of wood, was used about 4,000 years ago in Mesopotamia. In ancient Rome, locks more closely resembled those that we use today. They were made of iron, with an enclosed bolt.

Are all locks cylindrical?
No, there are three other types of locks: magnetic, combination and electronic.

How are magnetic locks made?
Small magnets are arranged on the key to this type of lock. When the key is inserted, the magnets align the pins and free the bolt. In electronic locks, instead, a card with a magnetic strip (similar to a credit card) is inserted in a slot. If the card is recognized by the scanner, the door is opened electronically.

What are combination locks?

Combination locks don't need keys. The earliest type was a series of blocks or rings, arranged side by side. There are slots on the inside edge of each one, with letters or numbers on the outside. When the exact sequence of letters or numbers is formed, the slots line up, freeing the pin that has been blocking them.

Do those with dials work in the same way?

The mechanism is essentially the same, but the blocks are inside the lock and only one numbered dial is turned to align them. The disk-shaped blocks, usually three, are held in place on a common axle behind the dial. Through a system of pins that can reach and turn the next block in line, the rotations of the dial will align the slots on top of each block, freeing the lock.

Are there many combinations?

With three disks, each of which may have a number from 1 to 100, there are a million possible combinations.

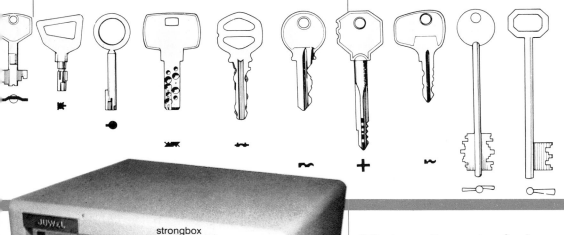

What are the parts of a key called?

The bow or "handle" is the part used to hold the key. The shoulder is the small jut on either side of the key, next to the bow, which determines the depth to which the key may be inserted in the lock. The blade is the long section of the key that is inserted in the lock. The serrations, also called cuts or bites, are the jagged edges that are cut to fit a particular lock. The warding is the cut on the flat side of the key that runs the entire length of the blade, serving to guide the key into the correct position in the lock.

Do safes also have locks with keys?

They did, up to the nineteenth century. Nowadays they have combination locks with dials which may even be regulated by clocks inside the safes called "time locks." With this system, the combination can open the safe only at set times.

strongbox

door handle blocking system

expansion bolt

THE ELECTRICAL SYSTEM

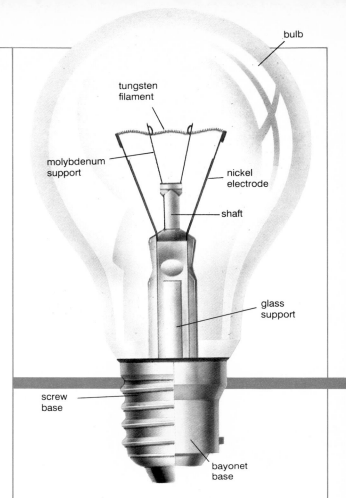

bulb

tungsten filament

molybdenum support

nickel electrode

shaft

glass support

screw base

bayonet base

How does a light bulb work?
The ordinary electric light bulb, like the one shown on the right, has a glass bulb in which a glass support holds a network of wires. At the end of this network, there is one wire, the tungsten filament, which can withstand very high temperatures. When the electric current enters the light bulb, it passes through the wires and, in particular, through the tungsten filament which heats up to the point that it emits very intense white light.

Doesn't the tungsten filament burn?
If oxygen were in the bulb, it would surely burn since it reaches a very high temperature. For this reason, before the bulb is sealed, it is emptied of oxygen and filled with a special gas.

How is electricity produced?
Electricity is a form of energy. Energy is neither created nor destroyed; it simply changes form. Electrical energy, an especially important kind of energy, can be obtained in several ways. On an industrial scale, it is produced in power stations by transforming the energy produced by the movement of river water (hydroelectric power stations), by the chemical reaction of petroleum (thermoelectric power stations) or by atomic fission (nuclear power stations).

Where does it go when it leaves the power stations?
From the power stations, the electric current is introduced into a transmission network which brings it to substations which convert the current from hign-voltage to low-voltage for home and business use. The distribution networks branch off from these substations and lead to homes, industries and other places that use electricity.

What happens when electricity arrives at a house?
The electric cable which enters a building from the distribution network usually ends in a central panel on which are found meters and circuit breakers. From the meters the cables go to individual apartments where they branch off, from a specially designed box, into a series of other wires which go to the sockets in each room.

Are there many kinds of sockets?
Yes, there are many kinds of sockets and plugs, depending on the power required and on the security measures needed.

What is a grounded connection?
It's a wire, buried deep in the ground and connected to water or gas tubing. It discharges dispersed current which would otherwise be dangerous. It is a safety device that must be incorporated into each installation.

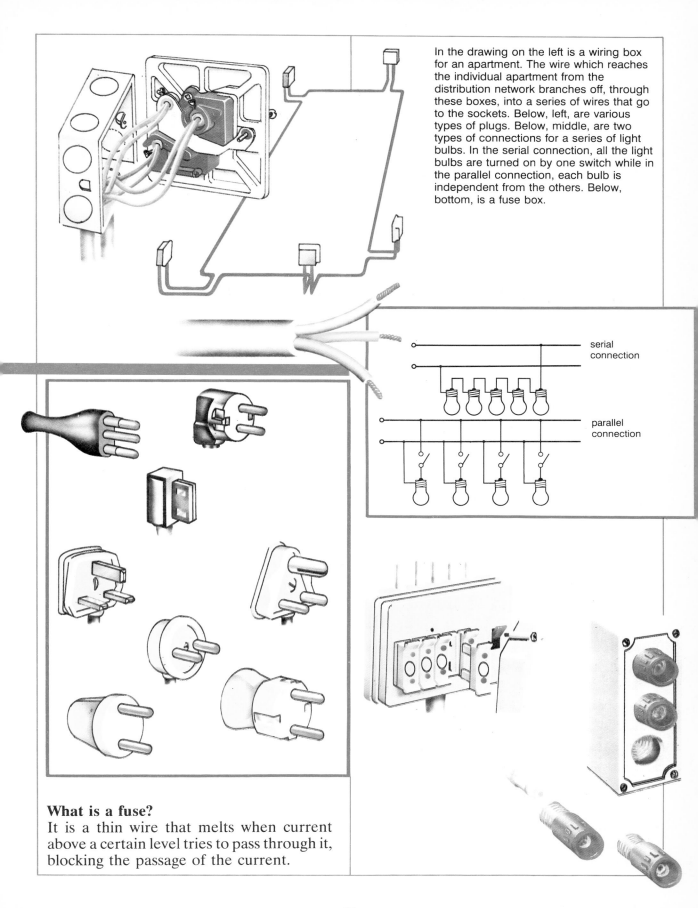

In the drawing on the left is a wiring box for an apartment. The wire which reaches the individual apartment from the distribution network branches off, through these boxes, into a series of wires that go to the sockets. Below, left, are various types of plugs. Below, middle, are two types of connections for a series of light bulbs. In the serial connection, all the light bulbs are turned on by one switch while in the parallel connection, each bulb is independent from the others. Below, bottom, is a fuse box.

serial connection

parallel connection

What is a fuse?
It is a thin wire that melts when current above a certain level tries to pass through it, blocking the passage of the current.

HEATING SYSTEMS

Are there different types of heating systems?

Yes, there are. The most common centralized heating systems circulate water heated by a furnace through radiators. But radiant panels or air systems may also be installed, especially in offices.

How do radiators work?

A furnace, which is normally found in the basement of a building, heats water which rises under pressure through the duct (the ascending column) and arrives in the radiators. The radiators, made of steel or cast iron, transmit heat to the room.

What happens to the water?

The water, which at this point is cool, having given off its heat to the radiators, descends into the furnace through another duct (the descending column), and the cycle begins again.

How is the water in the furnace heated?

It is heated by burning some type of fuel, such as methane, diesel or coal.

What is the burner?

It is the principal part of a furnace, the device which brings methane or gas in contact with the air to make a combustible mixture. It also ignites this mixture. The flame heats the water in the pipes.

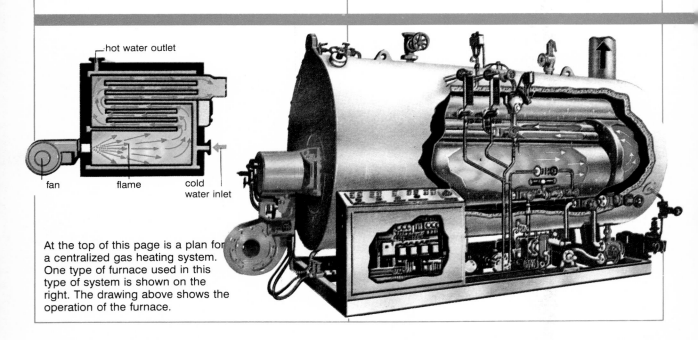

At the top of this page is a plan for a centralized gas heating system. One type of furnace used in this type of system is shown on the right. The drawing above shows the operation of the furnace.

How does a radiant heating system work?

The hot water runs, not in radiators, but in coils, under the floor of the house. As in radiators, the water in the coils releases its heat to the surroundings. This system is more economical than the use of radiators, but it makes the air very dry and, therefore, not very healthy. Besides, it is difficult to adjust the system in order to have an even and constant temperature in all rooms of the building.

How does gas reach our homes?

The gas used in homes is usually methane, which is found underground, or the so-called city gas or lighting gas, which comes from distilling coal. Methane is conveyed to places where it is used by large piping systems, which may be hundreds or thousands of miles long. City gas is stored in large gasometers, similar to cisterns. From the methane ducts or the gasometers, a system of pipes conveys the gas to homes. This distribution network is managed by special companies, usually municipal firms.

What is a meter used for?

Gas and water enter our homes through a meter which measures the amount we use. It has a tap that can turn off the flow, for example, when there is no one at home. Eliminating the flow from the meter eliminates the possibility of leaks within the house.

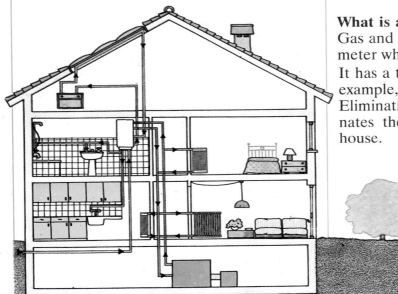

This is a diagram of a house with a solar heat system for heating water.
The panels on the roof absorb the heat of the sun and release it to the water that runs through them. The water is then sent to radiators and water outlets.

What about the systems that use air?

In air systems, the burners installed in the basements of buildings heat air instead of water. The air is then distributed to different rooms by ducts.

Are stoves still used to produce heat?

In the past, wood-burning and coal-burning stoves were used, and more recently, we have seen gas-burning stoves. Before that, wood and coal were burned directly on the hearth. Nowadays, centralized heating systems, such as those that we have described, are common. Hearths or fireplaces are rarely used except in mountain shelters or, in many cases, for decoration. Stoves are only used in areas lacking gas lines.

Are there also solar heat systems?

Yes, they are especially used for heating water for domestic use. Since gas heating is expensive and gas is a resource which, sooner or later, will run out, the idea of solar collectors is spreading. These collectors are panels, usually placed on the roof, which accumulate solar radiation, for example, in the form of electrical energy. This energy is transmitted to a furnace in the house. The panels may also have metallic surfaces which, absorbing the rays of the sun, heat up and transmit the heat to the water circulating in them. They may also be made of insulated glass which prevents the radiation from dispersing by reflecting it on metallic surfaces.

THE WATER SYSTEM

How does water reach our homes?

The water in our houses comes from rivers, lakes or wells. From these places, it is transported to a central plant where it is purified and disinfected. Afterward, it is piped to cities and towns and distributed to houses. Part of the water is stored in large reservoirs for use in case of drought. When needed, it is taken from the reservoirs by a system of pumps.

What are these pipe systems called?

They are called aquaducts and are used for the transport, distribution and regulation of water.

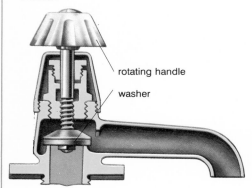

rotating handle

washer

How does a faucet work?

Faucets are devices which are inserted into water pipes. They are basically composed of two elements: a cable connected to the pipe and an element for adjusting or stopping the flow of water. Above you see a faucet with a handle which, when rotated, turns a screw which raises or lowers the element. This element, shaped like a truncated cone, blocks the passage of water. As this element is raised, the water in the faucet increases. The handle may have different shapes. Instead of a rotating handle, there may be a variety of systems using levers. Whatever system is used, it controls the opening and closing of a passage for the movement of water under pressure from the pipes to the sinks, bathtubs and other outlets.

What is "soil" water?

The water that arrives at our homes is conveyed by a single system of pipes. The waste water, instead, is brought to the sewers by two different systems, depending on the use it has had. We call the water which drains out of sinks and basins "white"; "soil" water, instead, is discharged from other facilities, including the toilet, and contains solid waste matter.

Where do sewers go?

They discharge their wastes into rivers or seas. People are, however, using many chemical substances, such as detergents which are poisonous to living organisms. Because of this, we must find alternative dumping grounds.

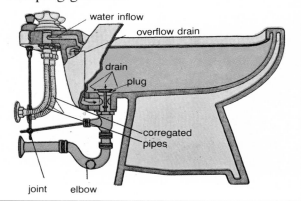

water inflow

overflow drain

drain

plug

corregated pipes

joint   elbow

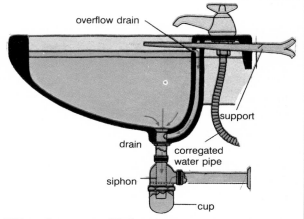

overflow drain

support

drain

corregated water pipe

siphon

cup

What happens if the sewers transport noxious substances?

Certain species of animals, for example, fish, may die. Some plant species may over-develop which, in the long run, can also be dangerous.

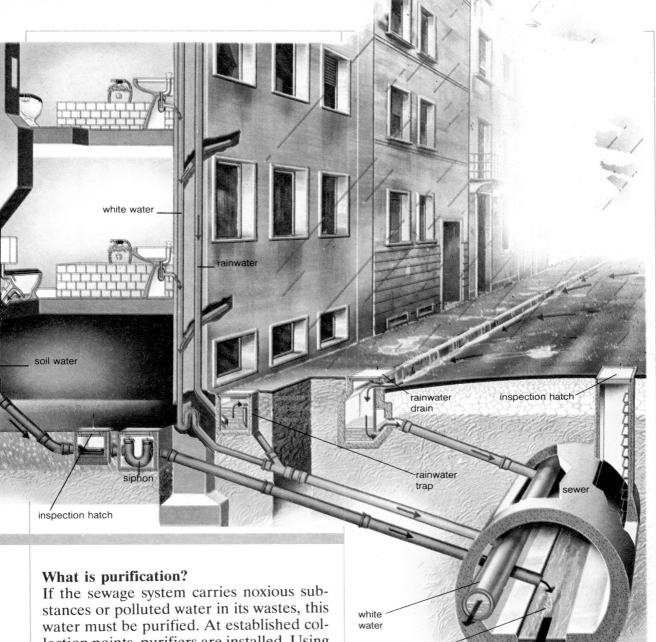

white water

rainwater

soil water

siphon

inspection hatch

rainwater
drain

inspection hatch

rainwater
trap

sewer

white
water

soil
water

What is purification?
If the sewage system carries noxious substances or polluted water in its wastes, this
water must be purified. At established collection points, purifiers are installed. Using
systems of filters, they treat the water and
then destroy the pollutants in it.

Does rain water also go into the sewers?
Yes, it is collected along with the white
water from domestic use.

What is a siphon?
A siphon, also called a trap, is located below
every drain in the house. It prevents waste
water from coming into contact with clean
water.

How is it made?
It's the U-shaped pipe which you can see
under the sink. When you use the sink, the
water accumulates in the curve of the U,
making a kind of trap between the sink and
the waste system. In this way, the siphon
isolates harmful gases and unpleasant
odors which could spread from the
bathroom drain pipes. It also prevents
bacteria and parasites from entering the
house through the hydraulic system.

SECURITY SYSTEMS

What is a security system?
It is any system which controls a place or a process and gives off a signal in case of a dangerous situation or unauthorized activity. Both security systems and alarm systems alert us to an intrusion in the house. They can detect smoke or high temperatures, alerting us to fires and helping to avoid some fire hazards.

Are there many kinds of security systems?
Yes, and they differ in two principal ways: the method used to determine whether the undesirable situation is occurring and the method used to give the alarm.

Are many of the systems electronic?
Many security systems are based on electric circuits. In one case, a normally closed circuit will be opened, or broken, when an intrusion occurs. Other systems have open circuits that set off an alarm when they are closed. A photoelectric cell, for example, can send an invisible light ray from one side of a door to the other. The light ray keeps the electric circuit closed. If something goes through the door, the circuit opens, and this change activates another circuit which sets off an alarm, like a siren.

Are there also systems that react to pressure?
Yes, you can also use pressure systems, placed under floors or carpets, for example. They take advantage of the fact that the pressure makes the resistance vary in an given environment, allowing current to pass. Normally the resistance there will not allow the electric circuit to close. But when the pressure increases, for example, when someone enters, the resistance changes and the circuit closes, setting off the alarm.

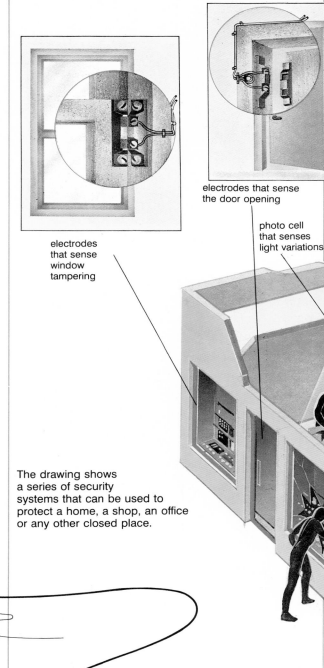

electrodes that sense window tampering

electrodes that sense the door opening

photo cell that senses light variations

The drawing shows a series of security systems that can be used to protect a home, a shop, an office or any other closed place.

Are radar and sonar also used in security systems?

Yes, radar and sonar are used, or in any case, we use systems based on the emission of radio or ultrasonic waves whose echo is registered. The echo changes when the signal encounters a moving object. In this way, these systems can show the presence of an intruder in places where there should be only inanimate objects.

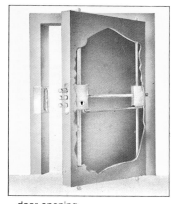

door opening which triggers alarm

closed-circuit TV

vibration sensor

close-circuit TV

infrared photo cells that give off an alarm when interrupted

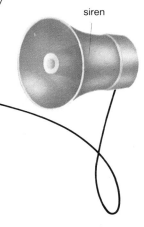

siren

Different kinds of alarm systems are available to protect cars from thieves. In the first example below, the current is prevented from reaching the ignition system. The second has a system of sensors which react to movement and set off a siren or horn.

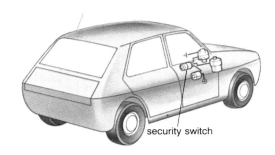

security switch

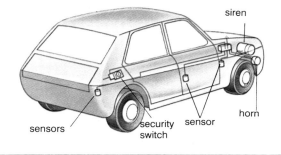

siren

sensors

security switch

sensor

horn

Do all security systems use acoustic alarms?

No, they are not necessarily used. In homes and on cars, sirens or similar devices are often used because they are easy and inexpensive to install. The alarm, however, may also be more complicated. It may turn on a warning light at a police station or in a guard room. In an industrial plant, there may be an electric signal which sets off an automatic emergency procedure.

DID YOU KNOW . . .?

Can the gas used in homes be dangerous?
Yes, it can because it contains carbon dioxide. A gas leak may be caused by a defect in the equipment, a mistake in using the gas or forgetfulness. It may poison the air or explode if there are lit cigarettes, matches or electric sparks in the area. A spark from a switch is even enough to cause an explosion.

Is gas visible?
No, and that's exactly the problem. Gas can expand rapidly without being seen. To make it recognizable, however, certain substances are added to it which give it a strong odor.

How does an electric meter work?
The user of electric energy must pay the producer for the amount of energy he has used. To establish this amount, we use a meter which basically consists of a small motor which moves a disk. The speed of the rotation is proportionate to the energy used. A small part of the current used passes through the meter. This amount is proportional to the total amount of current used beyond the point at which the meter is located. From the number of rotations of the disk, you can arrive at the amount of current used.

What is a watt?
It is the most common unit for measuring electrical power which is the amount of work done by electrical energy in a given amount of time. For practical reasons, we use the kilowatt as the unit of measure. The kilowatt equals 1,000 watts of power used continuously for an hour. The electric bill, which we periodically pay to the company that supplies electricity, is based upon the number of kilowatts that we have used and which have been registered by the meter in our house. To give one example, a normal refrigerator uses about one kilowatt per day.

What is the difference between direct and alternating current?
Electric current is nothing but a flow of electrons. In direct current, the flow is always in the same direction, from the positive to the negative pole. In alternating current, the flow periodically changes direction. The number of times that the flow changes direction in a second is the frequency of the alternating current. Batteries and generators produce direct current. The large generators at power stations usually produce alternating current, what we receive in our homes.

How does an electric stove work?
An electric stove has heating elements for each burner and for the oven. When you turn the dial, the current runs through the wire to the element and heats it. The element is surrounded by insulation which prevents it from becoming too hot.

What about the oven?
In an electric oven, there are two heating elements. One is placed on the bottom and is used for cooking. The other is at the top for broiling. These elements heat the air in the oven which, in turn, cooks the food. The broiling element heats up more intensely, and it gives a brown or golden color to food.

What is a thermostat?
By the word "thermostat," we mean any device that keeps the temperature at a preset level. To do this, the thermostat must have an element that is sensitive to temperature, which works like a thermometer. It must also have an element that can act on the heat source to turn it on or off or to adjust it. There are two types of thermostats. One is the "all-or-nothing" which turns off the heat when it reaches the desired temperature and turns it back on when the temperature goes down. The other, called "modulated action," works continuously, trying to maintain a constant temperature.

COMMUNICATION
AND INFORMATION

BOOK PRODUCTION

Is the production of a book complicated?
Yes, it is a process that has many different phases. The text and the illustrations follow different paths which come together in the phase that immediately precedes printing.

What is composition?
It is the process which involves the writing of the text. It is usually done with photo-composition machines.

Are there other composition methods?
The traditional ways which used movable characters set manually or mechanically are no longer used. Recently, instead, we have seen the rise of digital technology, in which the design of the characters is preserved in the memory of the photocompositor in a numeric form, like all other computer data. Transferring the text to paper usually occurs, then, with a procedure similar to photocopying.

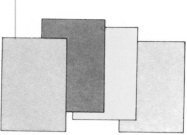

illustrations		photolitography	color separation	color proofs	films of text and illustrations combined
editing					
text		photocomposition	video text correction	proofreading	

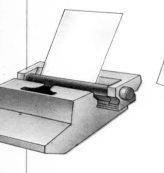
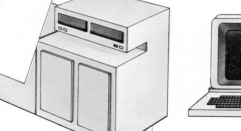

Why are they called photocompositors?
It is because the text is produced on photographic paper with an optical process. This machine has a terminal, similar to a computer, with a keyboard on which the text is typed. The machine controls different optical devices which etch the characters on photographic paper. For example, the characters may be etched on disks. The machine directs a band of light to the character you want on the disk, and the band projects the image of that character onto paper.

How are illustrations processed?
The illustrations, whether they are drawings or photographs, in black-and-white or color, go through a process called photolithography. In this process, the illustrations are put on negatives, or on several negatives in the case of colored illustrations, in the format needed for the particular publication. Not until the next phase are they usually joined with the text. The assembly of these two parts creates the printing plate.

Are there different ways of printing?

Yes, there are many printing techniques currently in use. The choice of one over another depends upon the kind of book you want to produce. The most common technique for books is offset printing while newspapers with a large circulation often use a process called rotogravure.

How are the matrices made for rotogravure printing?

They are made with a photoengraving process in which the matrix is actually engraved. At the end of the preparation process, there are many holes of varying depths, depending on the intensity of the color desired.

preparation of printing plate from film

printing

How is offset printing done?

In offset printing, the metallic plate, made from the negatives of the text and illustrations using a type of photographic process, does not come into direct contact with the paper to be printed. The plate deposits its ink on a rubber roller which transfers it to paper. If colored printing is done, every page is passed under four rubber rollers, each of which is inked with a different color, black or one of the three basic colors, red, yellow and blue.

How is rotogravure printing done?

The matrix engraved on the cylinder has holes of varying depths. The deeper holes collect more ink than the more shallow ones and, therefore, give a darker color. When the cylinder is put into the machine, it rotates in a tub of ink. A flexible blade on the surface of the cylinder wipes away the excess ink, leaving only the ink in the holes. The matrix, then, inks the paper against which it is pressed, and that makes the printed page.

PHOTOGRAPHY

How are photographs made?
The process of photography is based on the light-sensitivity of silver halides. Black-and-white film is coated on one side with a gelatin, called the emulsion, which has silver halides suspended in it. These crystals are modified by the amount of light that reaches them; more light creates more modifications. When you take the film out of the camera and put it in a developing solution, all the crystals which received the light become metallic silver. Where the light did not enter, the film is transparent. The image you receive is a negative, that is, the white parts are black and the black parts, white.

What do you do to get a positive print?
The paper for the final printing is also treated with silver halides. Light is passed through the negative onto this paper. Since this light can only pass through the transparent parts, the final image, by reversing the negative, is correct.

What about color photographs?
The procedure is similar, but the film has three layers of material, one sensitive to blue, one to green and the last to red. Depending on its color, the light reacts with the silver halides in one or more of these layers. All the other colors are obtained by combining different gradations of these three basic colors.

What is the objective?
In cameras, it is the system of lenses which collects the light and directs it to the film.

What is the shutter?
It is the device between the lens and the film which opens for an instant to expose the film. The time that the shutter remains open is called the exposure time; it may be set manually or automatically. The amount of light that reaches the film is also regulated by the diaphragm which restricts the opening of the lens to varying degrees.

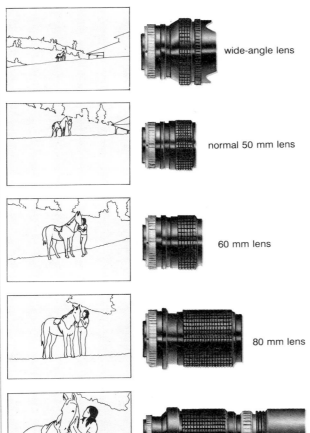

wide-angle lens

normal 50 mm lens

60 mm lens

80 mm lens

120 mm lens

250 mm telephoto lens

What does "reflex" mean?

There are two types of viewfinders. The direct one is simpler and allows you to see approximately what the lens sees. The other one, instead, allows you to see exactly what the lens sees; this is called a reflex viewfinder. What you see in the viewfinder is relayed by a mirror placed between the lens and the shutter.

What is the focal length?

If you look at an image through a magnifying glass, the image will be sharp only at a particular distance from the lens, which, in turn, depends on the size of the lens. That distance is the focal length of the lens. Camera lenses also have a focal length, and generally speaking, the image is proportional to the focal length of the lens. The greater the focal length, measured in millimeters, the closer the image on the film, as you can see in the sequence of drawings on the left. The more sophisticated cameras allow photographers to change lenses to take different kinds of photographs.

THE SLIDE PROJECTOR

What is a slide?
It is a positive photographic image impressed on a transparent film. For convenience, it is usually mounted in a rigid cardboard or plastic frame.

How does a slide projector work?
The slide is placed in front of a light source in the projector. The light goes through a condenser lens, through the slide and, finally, through a projection lens which relays the image to a screen. The condenser lens is used to increase the quantity of light that passes through the slide and to make the light rays parallel, which gives a sharper projected image.

What does the projection lens do?
It directs the light that has passed through the slide to the screen. It is focused by moving it backward or forward.

What kind of bulb is used in slide projectors?
Either incandescent bulbs, like those in a normal table lamp, or arc lamps are used. Incandescent bulbs give a good quality of light, but because they operate at high temperatures, require a fan. Arc lamps work by passing current through a gas, like xenon, whose atoms then give off light. Arc lamps give an intense light at a low temperature.

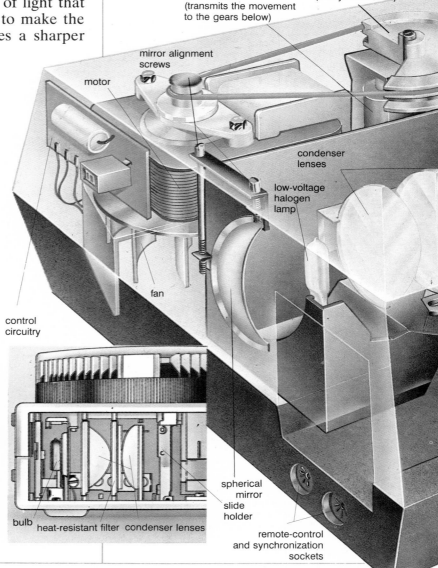

electromagnetic connection (transmits the movement to the gears below)

principal pulley (always in motion)

mirror alignment screws

motor

condenser lenses

low-voltage halogen lamp

fan

control circuitry

spherical mirror

slide holder

remote-control and synchronization sockets

Below is a model of a slide projector with a carousel, or circular, slide tray. The carousel is turned by remote control or automatically. When the slide reaches the viewing position, a device draws it out and puts it in front of the light for projection.

bulb   heat-resistant filter   condenser lenses

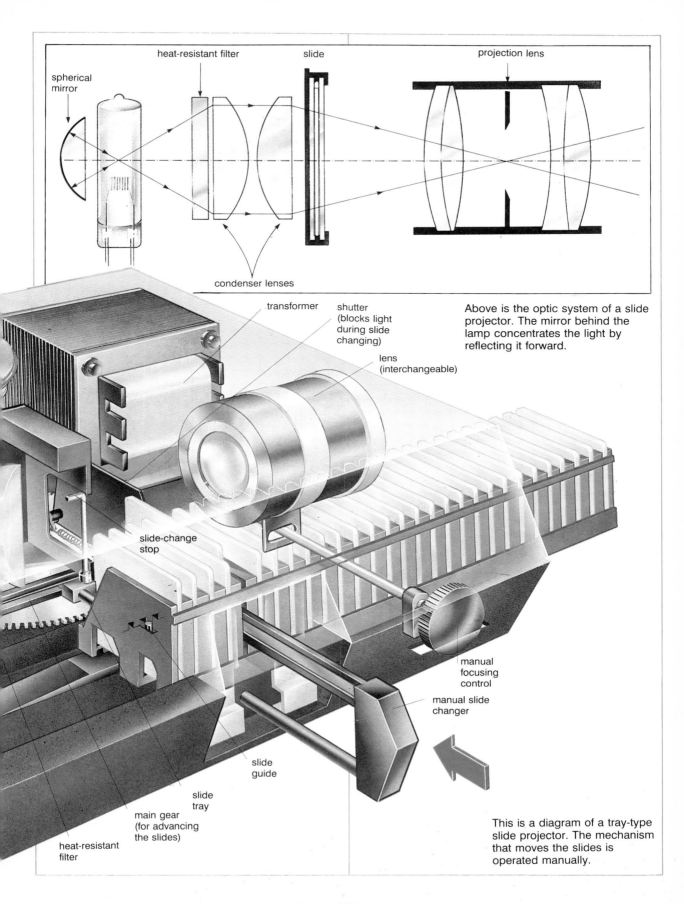

spherical mirror

heat-resistant filter

slide

projection lens

condenser lenses

transformer

shutter (blocks light during slide changing)

lens (interchangeable)

Above is the optic system of a slide projector. The mirror behind the lamp concentrates the light by reflecting it forward.

slide-change stop

manual focusing control

manual slide changer

slide guide

slide tray

main gear (for advancing the slides)

heat-resistant filter

This is a diagram of a tray-type slide projector. The mechanism that moves the slides is operated manually.

THE POLAROID® CAMERA

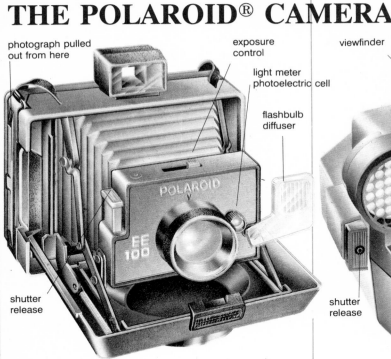

photograph pulled out from here

exposure control

light meter photoelectric cell

flashbulb diffuser

shutter release

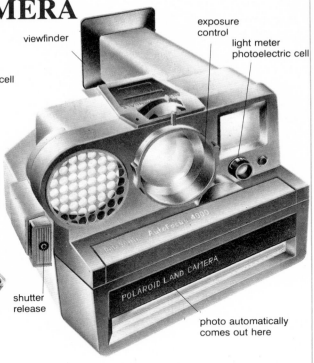

viewfinder

exposure control

light meter photoelectric cell

shutter release

photo automatically comes out here

What makes the Polaroid® camera special?

It is a special camera because it does the entire developing process. It not only produces a negative; it also produces a positive print. In the beginning, it only took black-and-white photos, but now it can take colored ones as well. The first Polaroid® camera was sold in 1948. It was invented by an American scientist, Edwin Land.

What is the name of the development technique used in this camera?

It is called diffusion-transfer development.

How does it work?

In black-and-white development, the negative paper is exposed, as in a traditional camera. Inside the camera, this paper is brought into contact with a gelatin-coated "positive" paper. The two sheets make a kind of sandwich, and when they are removed from the camera, they pass through a tight-fitting pair of rollers. The roller breaks tiny pods of developing jelly in the positive paper, and this jelly spreads evenly between the positive and negative sheets.

What produces the positive print?

The developing jelly reacts with the silver halide salts exposed to the light and forms a negative image. The silver halides that are not exposed actually move across the jelly and react with chemicals on the positive emulsion. Here they deposit grains of metallic silver and form a mirror image of the negative - that is, a positive print.

How long does this process take?

It usually takes from thirty to sixty seconds for the entire process to be completed.

Is the procedure for color prints more complicated?

Yes, the Polacolor® process uses a three-layer negative film and a four-layer positive part. The negative portion consists of three paired layers of silver halides and dyes. These layers, in order, are sensitive to blue, green and red light and, coupled with layers of dye molecules, give yellow, magenta and blue.

What happens when the negative film is exposed?

The blue-sensitive silver halides, in the first layer, become activated by blue light but not by any other color. The same is true of the next two layers of halides, those sensitive to green, which react only to green light, and those sensitive to red, which react only to red light.

What happens when you take out the film?

The same thing happens to this film, like black-and-white film; the pods of jelly break as they pass through the pair of rollers and spread the developing solution throughout the negative layers.

How does the development occur?

When the developer reacts with the activated blue-sensitive halide particles, it forces them to combine with the paired yellow dye molecules. The yellow, as a result, is trapped on the negative. But the molecules of the other two dyes can migrate up through the jelly-soaked layers of the negative to the positive part where they combine to form various shades of blue. This process, repeated for the three types of light at every point of the paper, produces a vivid mix of colors in the finished print. When the developer reaches the positive paper, there is a reaction which blocks any further development.

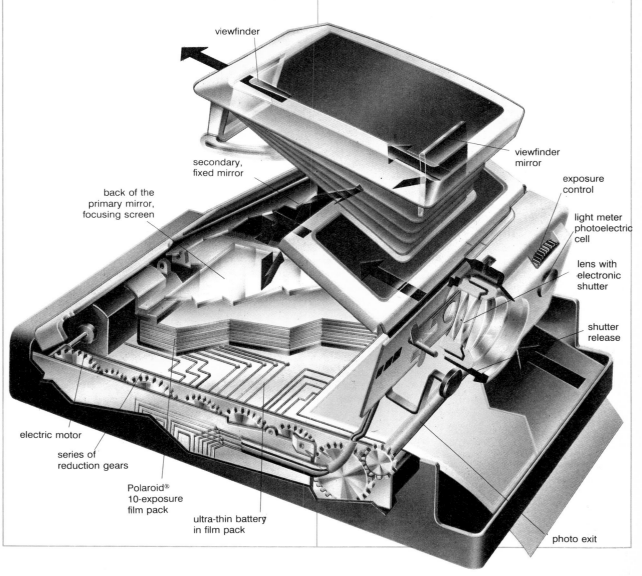

viewfinder

viewfinder mirror

secondary, fixed mirror

exposure control

back of the primary mirror, focusing screen

light meter photoelectric cell

lens with electronic shutter

shutter release

electric motor

series of reduction gears

Polaroid® 10-exposure film pack

ultra-thin battery in film pack

photo exit

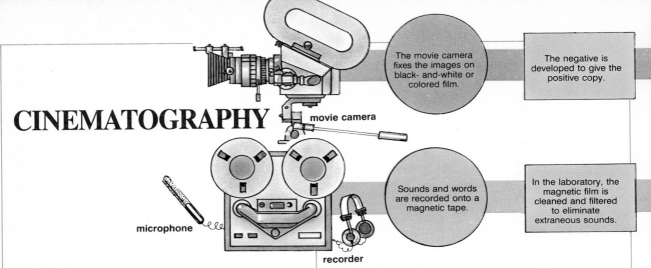

CINEMATOGRAPHY

movie camera

The movie camera fixes the images on black- and-white or colored film.

The negative is developed to give the positive copy.

microphone

recorder

Sounds and words are recorded onto a magnetic tape.

In the laboratory, the magnetic film is cleaned and filtered to eliminate extraneous sounds.

How is film for movies made?
It is like film for still photographs. On it are sequences of numerous photographs, all of them static images.

Why do movies seem to be moving?
During projection, the individual images follow each other so rapidly that our eye cannot separate them. This is due to a phenomena called image persistence. If the eye is presented with a series of images at a velocity of at least ten per second, the eye "attaches" one to the other as if it were seeing something happening. In fact, each image remains on the retina of our eyes for about one-tenth of a second.

Is the movie camera, then, like an ordinary camera?
Yes, it's basically like a camera for still photographs, but it can take twenty-four photographs a second.

What is a Moviola, or film-editing machine?
It is a special device on which an operator assembles the film, putting the different parts of the film in order, eliminating the extra or less successful parts, and recomposing it all into the final copy. In this phase, the sound, previously recorded on a tape, is also added. The operator must also synchronize the images and the sound, coordinating the images with the soundtrack.

How does a movie camera work?
Like a still camera, a movie camera has a lens that collects light from the scene being photographed and focuses it on a film. But unlike a still camera, a movie camera can capture action by taking pictures very quickly – moving the film forward and shooting at the rate of twenty-four frames per second. A pull-down mechanism draws the film from the film chamber into the gate, where the shutter opens, allowing light collected through the lens to hit the film and create images on it.

lens hood used to protect the lens and block extraneous light

variable focal-length lens

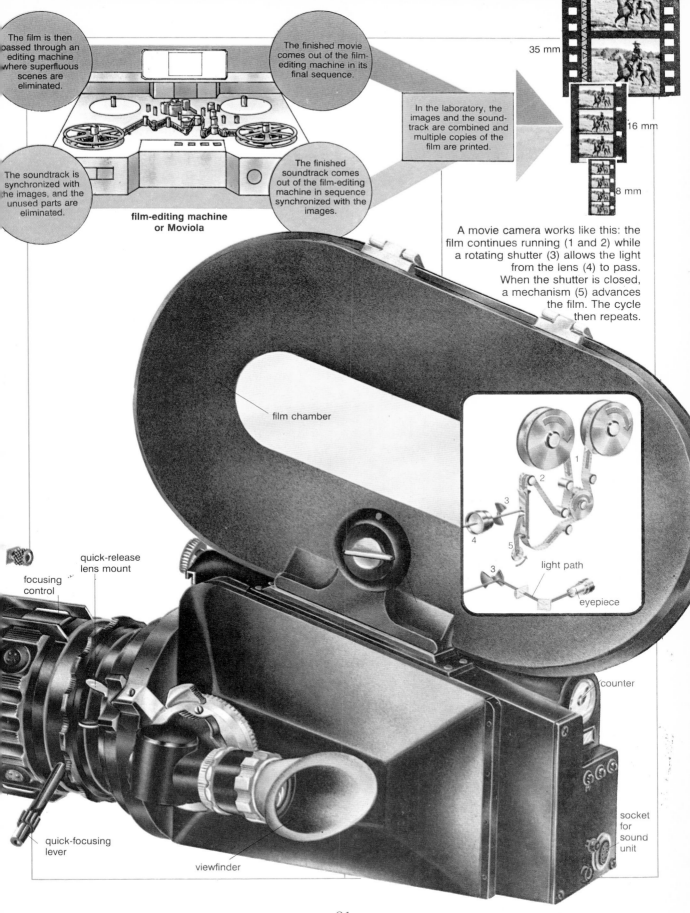

The film is then passed through an editing machine where superfluous scenes are eliminated.

The finished movie comes out of the film-editing machine in its final sequence.

In the laboratory, the images and the soundtrack are combined and multiple copies of the film are printed.

The soundtrack is synchronized with the images, and the unused parts are eliminated.

The finished soundtrack comes out of the film-editing machine in sequence synchronized with the images.

film-editing machine or Moviola

35 mm

16 mm

8 mm

A movie camera works like this: the film continues running (1 and 2) while a rotating shutter (3) allows the light from the lens (4) to pass. When the shutter is closed, a mechanism (5) advances the film. The cycle then repeats.

film chamber

light path

eyepiece

counter

quick-release lens mount

focusing control

socket for sound unit

quick-focusing lever

viewfinder

TELEVISION

Who invented television?

The first person who tried to transmit images was George Carey in 1875. His ideas were good, but the technology available at that time did not allow him to get good results. Paul Nipkow, an American inventor, was the first person to solve the problem of converting the images into signals that could be transmitted over a single wire. Modern television is largely based on the patents filed by V.K. Zworykin, a Hungarian working at the Radio Corporation of America, between 1923 and 1939. Television was made possible by the developments in electronics which occurred during the twentieth century.

How does a television camera work?

The basic function of a television camera is to send the light from an image onto a photoelectric plate. This plate has a very large number of photoelectric cells which produce an electric signal when they are stimulated by the light. The activity of each individual cell depends on the amount of light it receives. The cells on the plate are periodically explored by an electronic beam which surveys them line by line. Each cell discharges its electricity when the beam reaches it which makes it possible to record a series of electric signals which correspond to the intensity of the light that had stimulated each cell.

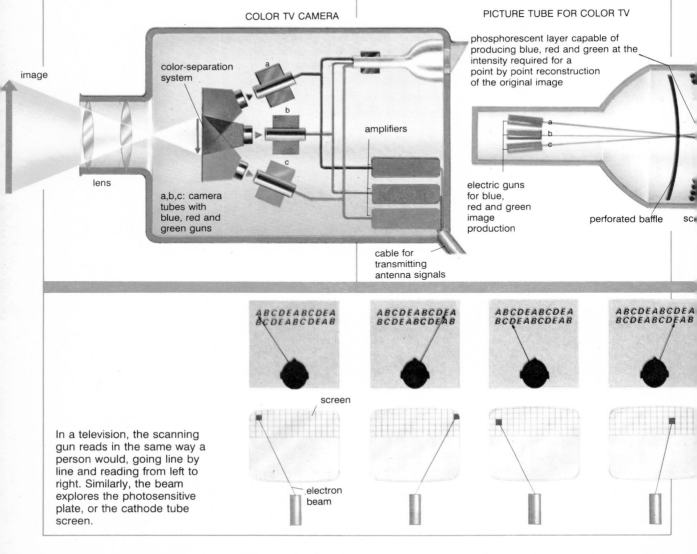

COLOR TV CAMERA

image

color-separation system

lens

a,b,c: camera tubes with blue, red and green guns

amplifiers

cable for transmitting antenna signals

PICTURE TUBE FOR COLOR TV

phosphorescent layer capable of producing blue, red and green at the intensity required for a point by point reconstruction of the original image

electric guns for blue, red and green image production

perforated baffle

screen

In a television, the scanning gun reads in the same way a person would, going line by line and reading from left to right. Similarly, the beam explores the photosensitive plate, or the cathode tube screen.

screen

electron beam

How are television images received?

The electric signals recorded by the electronic beam are first transmitted by radio waves. They are used to modulate the radio waves which are captured by a receiving antenna and transformed again into electric signals by conversion devices.

Does color television work the same way?

It works in a similar way, but, obviously, everything is a little more complicated. There are three electronic beams and three electronic guns that produce them, for blue, red and green. In the receiving apparatus, a deviation baffle sends the electronic beams to the appropriate photosensitive material.

How is the image reconstructed?

In a television, there is an electric beam similar to the one at the source. The incoming signals guide it while it explores the internal surfaces of a cathode tube on which are phosphorescent particles. The incoming signals tell the individual particles when there is enough energy to light up. The succession of cells stimulated in the photoelectric plate determines the design of the on-and-off elements on the television screen, reproducing the original image.

In a color television, the light rays from an image pass through a system of lenses, mirrors and filters which split them up into the colors red, green and blue. In the television camera, there are three tubes, one for each color. The electric signals which are transmitted by the sending antenna and which reach the receiving antenna, are a mixture of signals from the three tubes. In the receiving apparatus, there are also three electronic guns, as in the camera.

THE VIDEO RECORDER

What is a magnetic tape?

It is a fine band usually made of plastic and coated with an even thinner band of magnetizable metallic oxide particles, such as the ferrites. Since each electric signal also has a correspondingly intense magnetic field, electric signals of an auditory system like radio or a visual system like television can be recorded on a tape. The magnetic field generated by those signals modifies the distribution of the magnetic particles on the tape.

On the right is a portable video recorder and TV synchronizer, the part of the TV which receives the signal from the antenna and transforms it into electric signals.

Is this process the same for both tape recorders and video recorders?

Yes, all recorders register electric signals on the tape in a magnetic form, and the basic principles are the same. The difference lies in the fact that tape recorder signals are then used to control the speakers while video signals control the electronic guns of a TV set. They also control the TV speakers because the auditory band is recorded on the videotape as well.

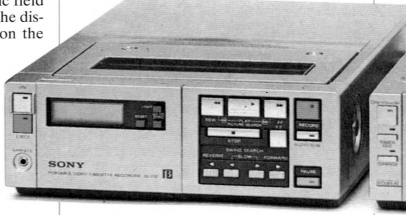

How is recording done?

When the electric signal is passed through a wire coil wrapped around a metal core, it will generate a magnetic field within the core. At one end of this core is the recording head, the outlet for the magnetic signal. If the tape is passed along this head, the magnetic field at any given point will cause the magnetized particles to rearrange themselves into a particular pattern. Once the tape has been altered in this way, it takes on an imprint of its own, a kind of magnetic signature of the original signal, or television image.

How is the signal reproduced?

Since the tape retains the distribution of the particles, the procedure can be reversed, using the magnetic signals generated by passing the tape near the head to reconstruct the electric signals.

On the right is a TV camera for videotaping. The diagram below shows the principal phases of video recording. The TV camera (1) is connected, by means of a control (2) and synchronization (3) group to a video recorder (4) with a color unit (5) or a cassette recorder (6). The microphone (7) records the sound. The color TV (8), equipped with sockets, can be used as a monitor.

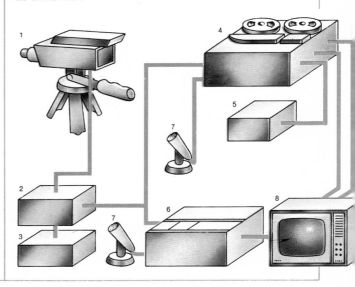

Is a TV camera for videotaping like a film camera?

They have a similar function but use a different procedure. The movie camera takes photographs; it is like a still camera. The TV camera also collects the light from the scene it is recording but uses it to generate electric signals and to record them in magnetic form on the tape. The procedures are therefore different, although the result is the same – to produce moving images.

Is the speed at which a tape runs important?

More important than the speed itself is the constancy of the speed which insures that the succession of electric signals will be precisely reproduced. All video recorders, like audio recorders which have a similar problem, have devices which insure that the tape runs at a constant speed. The usual speed is thirty-eight centimeters per second.

Can you videotape without a television?

The TV screen, in reality, is not used for taping. It is enough to have the part of the television that receives signals from the antenna and decodes them into electric signals. For this reason, many video recorders can tape with the television turned off or tuned to another channel. It is now possible to buy tuners which only receive TV signals but, having no cathode tube, do not transmit images.

THE VIDEODISC

What is a videodisc?

It is a disc, about the size of a 33 rpm record, on which are recorded visual images, similar to those found on a videocassette.

Does it work like a record?

No, they are very different. Videodiscs are not magnetic; they use laser-based optical technology. The information is recorded as tiny "pits" made by laser, and they are read by a laser beam in the videodisc player. This player is connected to a device that converts the information into electric signals which, in turn, control the cathode tube of a television.

How is information recorded on a videodisc?

This recording process is more complicated than the magnetic method. The information – the images and the related sounds – is first recorded on a videocassette (1 and 2 in the sequence on the opposite page). Then the electric signals generated from the reading of the cassette are used to modulate the frequency of the carrier wave (3).

What does "modulate the frequency" mean?

All waves, including electric and sound waves, have three characteristics: amplitude, frequency and shape. The amplitude is the maximum level that the wave reaches. The frequency is the number of times that the wave repeats in a given period of time. The shape is the movement, more or less simple, of the wave. Each of these characteristics may be held constant while the others are changed. In the case of frequency modulation, the amplitude and, in this case, the very simple shape, remain the same while the frequency is changed in respect to another signal. The variations in frequency, therefore, code the starting signal. If the signal modified in this way is then recorded, you can extract and reconstruct the original, with all of its original characteristics, by using a suitable decoder.

How do you get from the FM signal to the pit on the disc?

The only characteristic of the wave that changes is the distance between the crests of its waves. If you "cut" the signal with a continuous line, (4 in the sequence on the opposite page), between the crests and the bases, you will get a succession of longer and shorter dashes, separated where the signal crosses this imaginary line. The small incisions on the surface of the videodisc (5) reproduce these dashes, from which the modulated signal can be reconstructed.

Are the dashes made by laser?

Yes; on each side of the videodisc, there is room for about twenty-six billion. The layer with these pits is coated with a transparent protective layer (6 and 7).

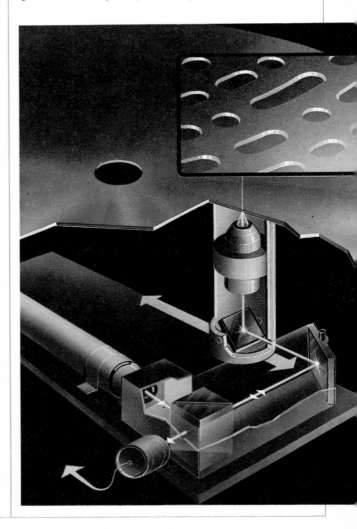

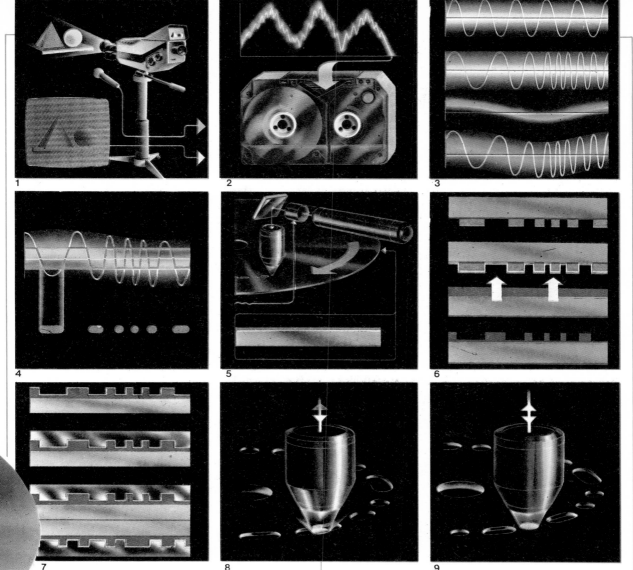

How is a videodisc read?

It is read with a laser beam (8 and 9 above) which is aimed at the surface of the disc as a tiny luminous spot from the lens. If this beam falls on a hole in the recording, the light does not return and the signal is lost. If it falls between one hole and another, the reflection sends back the light, transformed into an electric signal, which reproduces the FM signal. From this, special instruments reconstruct the original electrical signal.

At what speed do videodiscs revolve?

They are very much faster than records, spinning at thirty rpms a second or 1,800 rpms a minute. Each revolution corresponds to a single photograph.

Is this the only type of videodisc?

No, there are many other types of videodiscs. The one described here, however, is the most common. Another system has ridges on the surface like a record and is read with a stylus. Yet another uses the pits on the surface of the disc but reads them in a different way, using a stylus. A small amount of electric current is passed over the disc and the stylus reveals the differences in electricity between the parts that have pits and the parts that don't.

THE TELEPHONE

Who invented the telephone?
The first telephonic experiments were done in the nineteenth century by an Italian immigrant to America, Antonio Meucci. But the first person to succeed in commercially exploiting the invention was an American, Alexander Graham Bell, who patented the telephone in March, 1876.

How does the telephone work?
When you speak into a telephone receiver, the sounds act on carbon granules which are stored behind a thin dividing wall, the membrane. When the vibrations from your vocal cords strike the membrane, it begins to vibrate, which causes the carbon granules to compress and decompress alternatively. When the granules are compressed, they offer less resistance to electric current. When they are more compact, they give greater resistance to the passage of the current. Your voice, coming out of the telephone, then, produces a modulated electric current which travels along the telephone lines to a receiving apparatus.

What happens in the receiving apparatus?
When the modulated electric current reaches the opposite end of the line, the inverse process occurs, converting the electrical impulses into sound waves which, in turn, can be perceived by the person who is listening.

What is sidetone?
It is the resonance in your voice when you speak into a microphone. Sidetone, which could disturb communication, is eliminated by special wiring that blocks your voice from your receiver and allows it to be transmitted to the person you have called.

What is a microtelephone?
"Microtelephone" is the technical word for the receiver. It has two parts, a microphone and a receiver, which, in modern models, are placed in the most suitable positions for the user's mouth and ear. The part with the microphone has a membrane and a small cup for the carbon granules.

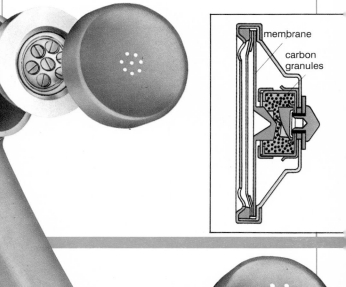

membrane

carbon granules

What happens when you use the telephone?
When you pick up the receiver, electric current passes through the line to the local telephone exchange. The flow of current acts as a signal to the exchange that a new call is coming in. The dial tone shows that the exchange is ready to accept your call. As you release the dial after each digit of the number you are calling, the flow of current is turned on and off from one to ten times, depending on the digit you have dialed.

How does a call reach the person you have telephoned?

The first telephonic connections were direct: a wire connected one user to another. Nowadays, a system like that would be unthinkable, since there are so many users. All telephones are connected to switching exchanges in which the calls are shunted automatically according to the telephone numbers of the users. The numbers that you call, whether by dialing or by pressing buttons, show the equipment in your switching exchange which other exchange to connect you with. The numbers also indicate which circuits within that exchange must be activated to allow you to speak with the person you have called. This process is entirely done with electromechanical or electronic equipment.

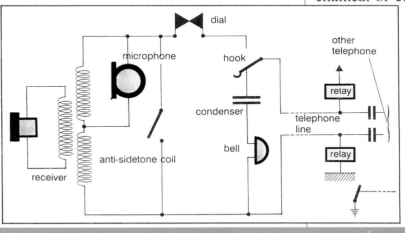

A push-button telephone works like a traditional rotary telephone. The only difference has to do with the buttons. Each time you press one of the numbered buttons, a device in the wiring produces two out of a possible eight tones, which you can hear while you are dialing, and converts them into electrical impulses which pass through the wires to your local telephone exchange. These tones act as a code for each digit and take the place of the electric signals of the traditional dial. In the diagram above is a complete telephone circuit, while on the opposite page is the microphone and receiver apparatus, including the anti-sidetone coil.

MUSICAL INSTRUMENTS

How does an electric guitar work?

The basic principle is the same as that of a normal, acoustic guitar: the vibrations of the strings produce sound waves, that is, waves of pressure in the air. However, instead of being amplified by a sound box, these waves are recorded by a pickup, a device that is quite similar to the controls of a tape recorder, in the body of the guitar itself. The pickup is a series of electric coils which produce a magnetic field. The sound waves produced by the guitar strings alter the magnetic field and are transformed into variations of electric signals. These signals can be sent to an amplifier and, from there, to a speaker.

Why would you use an electric rather than an acoustic guitar?

Above all, you can regulate the volume of the sound produced. This is an advantage because the sound of an acoustic guitar is rather weak and could not be heard in an orchestra above the other instruments like the winds, such as a trumpet or saxophone. Secondly, it offers the possibility of "treating" the electric signal in order to get different sounds. Apart from simply regulating tones, to vary their dryness or richness, there is also equipment which produces special effects.

Are there other electric instruments?

Using principles similar to that of the guitar pickup, electric models have been made of other instruments, like the violin and the piano.

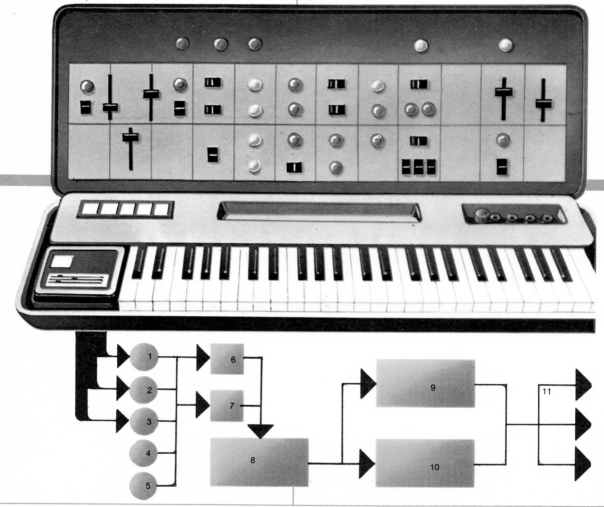

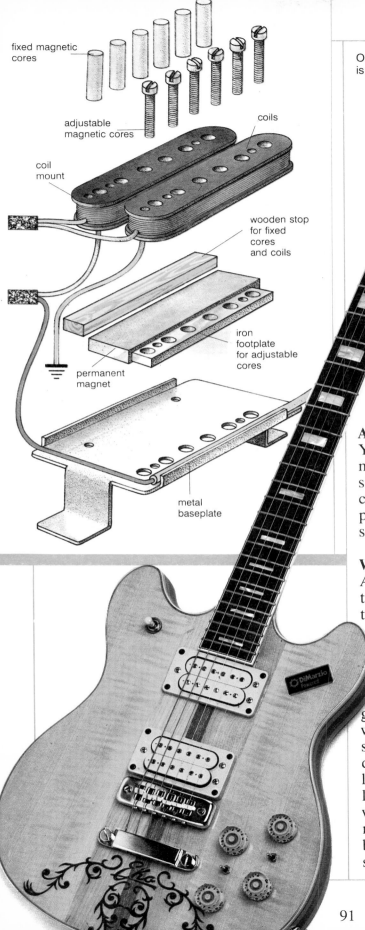

fixed magnetic cores

adjustable magnetic cores

coils

coil mount

wooden stop for fixed cores and coils

iron footplate for adjustable cores

permanent magnet

metal baseplate

On the left you can see how a pickup for an electric guitar is made.

What is a synthesizer?

In an electric guitar, the sound is produced by making the strings vibrate. The sound is then elaborated and amplified electrically. Nowadays there is also equipment that generates the sound directly by electronic means, using oscillating electric circuits or systems that are purely numerical. All these instruments are generically called synthesizers.

Are there many kinds of synthesizers?

Yes, since 1964 when an American engineer, Robert Moog, made the first synthesizer, many kinds of synthesizers have been constructed. They differ from one another principally in the system used to generate sound, which may be analogic or digital.

What does "analogic" mean?

Analogic means any device based on quantities that vary in relation to other quantities. For example, a mercury thermometer is an analogic instrument because the height of the column of mercury varies in relation to the temperature. In analogic synthesizers, the sounds are produced by oscillating circuits which generate continuous electric waves. These waves reproduce the movement of the sound waves. On the opposite page is a diagram of a classic analogic synthesizer, like those made by Robert Moog. The oscillators (1-5) produce undulating signals which are then processed by the wave-form modifiers (6 and 7) and filters (8) before being amplified (9 and 10) and sent to the speakers (11).

THE COMPUTER

What does "digital" mean?

It means numeric. It describes anything which uses discrete quantities, like numbers, rather than continuous quantities as in analogic systems. Taking the example of musical synthesizers, which were described in the previous article, the digital ones represent the musical waves as numbers, or more precisely, with values that correspond to particular points of the wave. These values are numerous enough to allow for the reconstruction of the wave that they represent.

What is the most common digital application?

It is, without a doubt, the modern computer, which is based entirely on digital circuits. The basic elements of a computer have only two states, on or off, which are usually indicated by the numbers 0 and 1. The operation of a computer is based entirely on the manipulation of 0s and 1s, following the rules of the binary numeric system.

What is the binary system?

It is a numbering system, similar to the decimal system that we use everyday. But instead of being based on powers of ten, it is based on powers of two, and, therefore, uses only two numbers. For example, in the binary system 11 is equivalent to three, not eleven. It is calculated like this: 1 times 2 to the first power (2) plus 1 times 2 to the zero power (1). In the decimal system, instead, it is 1 times 10 to the first power (10) plus 1 times 10 to the zero power (1).

How is it possible for a computer to do so many different things?

With the binary system, you can represent all the numbers as well as the letters of the alphabet including words and sentences. It is possible to do operations on numbers that correspond to operations on letters or words. The computer has a very large number of elements and can perform operations on numbers very quickly.

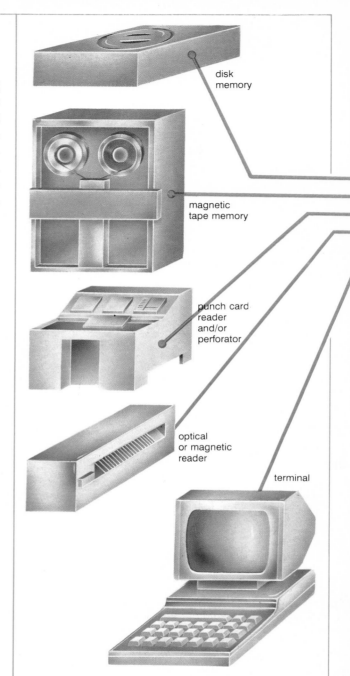

disk memory

magnetic tape memory

punch card reader and/or perforator

optical or magnetic reader

terminal

How is a computer made?

Computers essentially have three working parts. There is at least one processing unit in which all the operations are carried out. There is a memory unit for the programs, the data to process, and the processing results. Then there are input and output units (like the keyboard and the video) which allow the user to communicate with the processing unit and the memory.

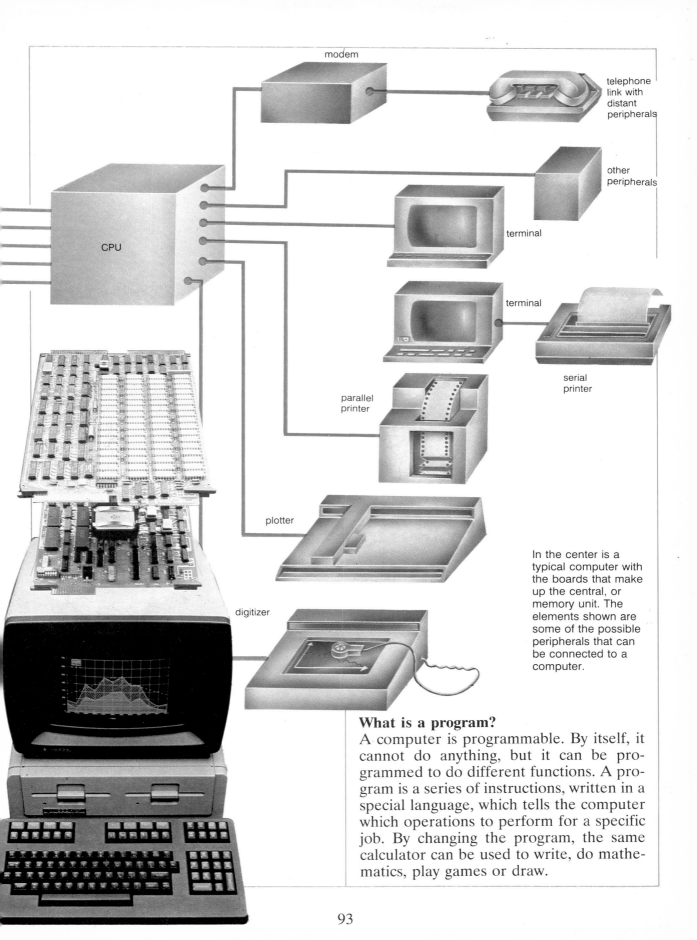

modem

telephone link with distant peripherals

other peripherals

terminal

CPU

terminal

serial printer

parallel printer

plotter

In the center is a typical computer with the boards that make up the central, or memory unit. The elements shown are some of the possible peripherals that can be connected to a computer.

digitizer

What is a program?
A computer is programmable. By itself, it cannot do anything, but it can be programmed to do different functions. A program is a series of instructions, written in a special language, which tells the computer which operations to perform for a specific job. By changing the program, the same calculator can be used to write, do mathematics, play games or draw.

COMPUTER NETWORKS

How are computer programs and information stored?

All computers are equipped with external memory devices, on which programs and information are saved. The most common device is the plastic floppy disk on which all the data are written magnetically, as on a record, but always as a succession of 0s and 1s. There are also metallic hard disks, which are sealed in closed containers and can hold a large amount of data. Large computers often store data on magnetic tapes. Yet another type of external memory is the magnetic card which resembles a credit card.

How much data can you put on a disk?

Hundreds of thousands of characters, or equivalent data, can fit on a floppy disk while a hard disk can hold tens of millions of characters. Tapes can hold hundreds of millions, or even billions, of characters.

What is the difference between one computer and another?

All computers are based on the same principles, and work, more or less, in the same way. They may differ in size, the largest computers being the most powerful. Or they may differ in the processing unit, with different processing units able to do different elementary operations. Since these operations are really very simple ones, like finding the sum of two numbers, even computers with different processing units can perform the same functions. The programs, however, will change because each objective requires a different string of operations.

What is an operating system?

It is the principal program of the computer, the one that allows the computer to manage all the basic information and program traffic. All programs use the operating system for certain basic processing, for example, turning on the external memory or the terminal. For this reason, the programmer doesn't have to rewrite these kinds of instructions for each program.

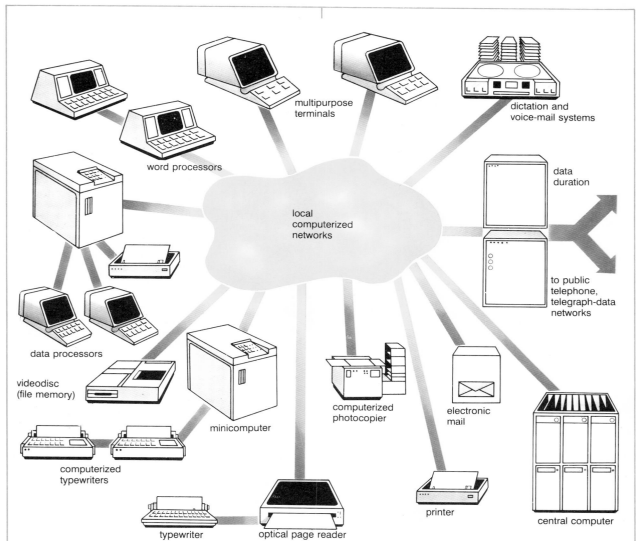

Diagram labels:
- multipurpose terminals
- dictation and voice-mail systems
- word processors
- local computerized networks
- data duration
- to public telephone, telegraph-data networks
- data processors
- videodisc (file memory)
- minicomputer
- computerized typewriters
- typewriter
- optical page reader
- computerized photocopier
- electronic mail
- printer
- central computer

What is a computer network?

It is a system that connects computers and related equipment. (See the figure above for an example.) The networks may be local, for connections within a single office or building. Or they may be geographic, for long-distance connections between cities, countries or even continents.

Do networks include many computers?

Two or three computers are enough to create a network. But networks may also have dozens, or even hundreds or thousands, of computers. The machines connected to a network don't have to be computers; they may be, for example, printers, special typewriters, terminals or memory units.

Why have computer networks been created?

Many people may work in an office. If all of them have a computer, it may be useful for them to communicate with one another using their computers instead of written documents. Besides, it is generally a good idea to have all the important office information in one place, like a central computer, which the employees can access through their own terminals. It may be practical to share equipment, for example, a very expensive printer or a memory unit with a large capacity. A company can also be connected with its branches and can receive long-distance orders, requests or data by means of a telephonically-connected computer.

COMPUTER GRAPHICS

Can images also be processed on a computer?
Yes, a good terminal is needed, but the computer can handle images. It treats them, like any other kind of data, as numbers. An image is considered to be a series of points, for example, black and white points. What is saved in the memory of the computer is the sequence of points that are either turned on or off. For a black-and-white picture, a sequence of 0s and 1s is sufficient: 0 for black points and 1 for white points. For color pictures, more numbers are used for each point (the exact number

Is the analogic-to-digital conversion used only for pictures?
No, it has general application whenever we want to communicate a continuous quantity to the computer, which understands only numeric or digital quantities. To take another example, sounds can be sampled and quantized for storage in the memory of the computer, after which they can be processed.

How good is this conversion process?
That depends on the sampling frequency (the size of the squares considered, for an image) and the quantization resolution (how large the range of usable numbers is; for an image, the number of colors that can be distinguished).

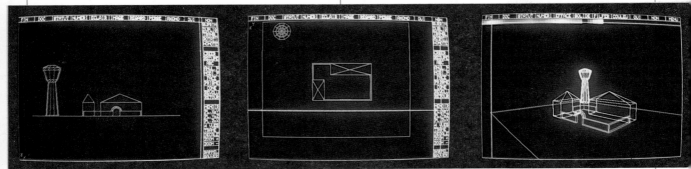

depends on the number of colors needed), but the principle remains the same.

Does the computer create the images?
The images can be created directly by the computer, or they can be "imported" into the computer with special equipment which "reads" a finished picture or photograph and transforms it into a numeric version.

How do you import an image?
The main device used is a analogic-to-digital converter. Let's suppose that the original image is a photograph. It is an analogic, continuous, image. The converter explores it and subdivides it into a lot of tiny squares. This process is called "sampling." For each square, it establishes an average color value and expresses it numerically, a process which is called "quantization."

How are pictures created at the computer?
As in the case of imported pictures, you must tell the computer which points to consider and which color they must have. You do this by writing programs or by using drawing instruments like light pens or drawing boards along with specific programs for them.

The illustration at the bottom of the opposite page shows a group of devices for processing images: the computer (1) with its disk memory unit (2), a graphics processor (3), which transforms numeric images into video signals, a color graphics screen (4), a screen for programs and messages (5), the keyboard (6), a drawing board (7), with a light pen (8), which can be used for free drawing, and a scanner (9), an analogic-digital converter. At the top is a computer-made design. In the center is a sequence of three-dimensional views of houses, from the front, from above and from various angles.

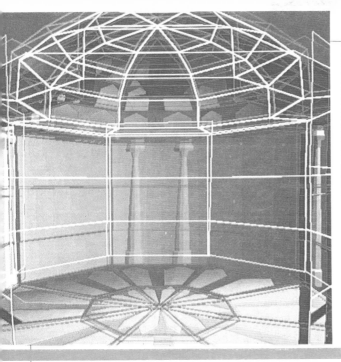

What kind of processing can be done on images?

That depends upon the program. There are programs that allow you to change colors and forms. There are others that can give different perspectives of a three-dimensional image. These programs are especially useful for architects who can explore different ways of constructing the same building and can see very quickly what impression the building would make from different angles or in different environments. There are also programs that allow you to create animation, a succession of static analogic images of animated cartoons. Computer animation programs give much more realistic results than traditional animated cartoons.

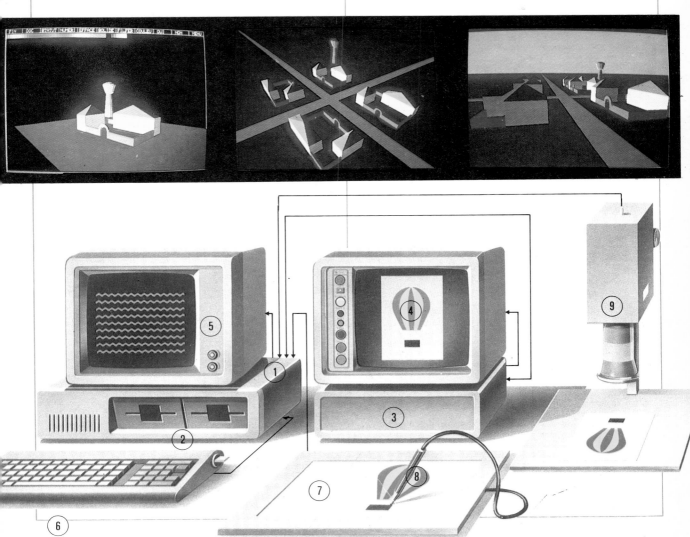

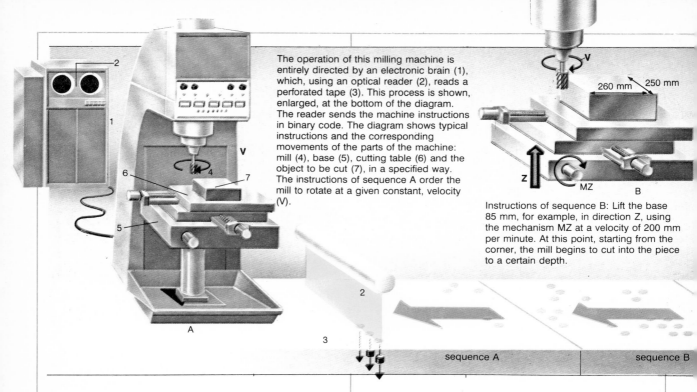

The operation of this milling machine is entirely directed by an electronic brain (1), which, using an optical reader (2), reads a perforated tape (3). This process is shown, enlarged, at the bottom of the diagram. The reader sends the machine instructions in binary code. The diagram shows typical instructions and the corresponding movements of the parts of the machine: mill (4), base (5), cutting table (6) and the object to be cut (7), in a specified way. The instructions of sequence A order the mill to rotate at a given constant, velocity (V).

Instructions of sequence B: Lift the base 85 mm, for example, in direction Z, using the mechanism MZ at a velocity of 200 mm per minute. At this point, starting from the corner, the mill begins to cut into the piece to a certain depth.

sequence A | sequence B

AUTOMATION

What is automation?
Any process which is done without human intervention is called automatic. Automation is the use of automatic processes, whatever the method chosen.

Is the computer used in automation?
Yes, the computer has a fundamental role in automation, particularly in industrial automation. A computer can be connected to sensors, devices which record particular quantities, like temperature, pressure, sounds or images. It can also be connected to control elements which can take actions, such as closing valves, turning on pumps, moving or starting mechanisms, and so on.

What is the role of the computer?
The computer plays the principal role. It receives information from the sensors, transformed into numerical data by analogic-to-digital converters whenever the quantity is measured analogically, as, for example, in the case of temperature. It evaluates the data, using its programs, and makes the necessary decisions based on the job to

be done, which, here again, is determined by its programs. It then activates the appropriate control elements. So, for example, in an industrial application, the computer may order the opening of the valves when the pressure reaches a certain level.

What are the advantages of automation?
Automated processes, in general, and, specifically, the use of the computer, offer numerous advantages. A computer can work at a constant level, without getting tired, without losing attention, and with great precision, twenty-four hours a day. For human beings, this level of constancy is impossible. Automation, therefore, is very useful in situations that require extreme precision and attention, or in boring and repetitive jobs. Computer control can also be very helpful in dangerous situations, or in places where human beings cannot go. The most obvious example are interplanetary explorations in which many tasks can be done automatically in places where humans, at least at the present level of technology, cannot go.

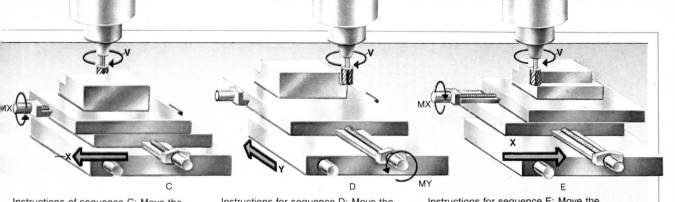

Instructions of sequence C: Move the cutting table 260 mm to the left in the direction - X, using the mechanism MZ at a velocity of 180 mm per minute. By the end of this operation, the mill will have cut a step along the first side of the piece.

Instructions for sequence D: Move the cutting table 250 mm in the direction Y, using the MY mechanism at a velocity of 180 mm per minute. As a result of this operation, another side of the piece will have a step cut.

Instructions for sequence E: Move the cutting table 260 mm to the right in the direction X, using the MX mechanism at a velocity of 180 mm per minute. After this operation, the third side will be cut and the work can continue with other instructions.

sequence C

sequence D

sequence E

Are these machines intelligent?

They are intelligent in the sense that they can do complex jobs which, up to now, have required human intervention. But being intelligent also implies being original and creative. These machines, instead, are limited to repeating, precisely and tirelessly, the same operations. Any possible variations must have already been programmed into them.

Above is the operation of an automated milling machine. Below is a large control room. The technology of automation is still in its infancy. We are trying to make machines still more automatic and we are looking for ways to facilitate communication between man and machine. In particular, we are experimenting with ways to make the machine understand the human voice.

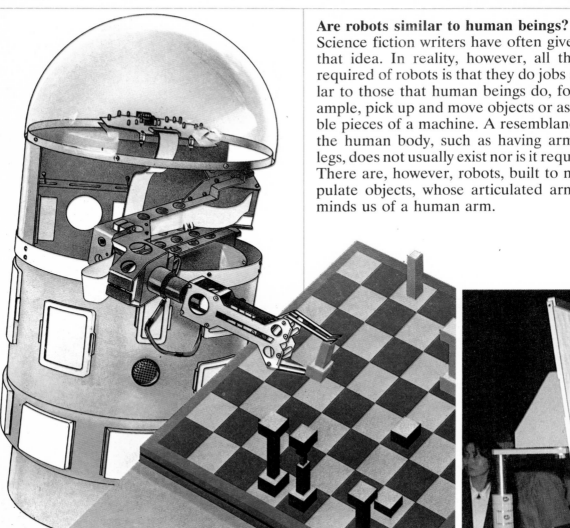

Are robots similar to human beings?

Science fiction writers have often given us that idea. In reality, however, all that is required of robots is that they do jobs similar to those that human beings do, for example, pick up and move objects or assemble pieces of a machine. A resemblance to the human body, such as having arms or legs, does not usually exist nor is it required. There are, however, robots, built to manipulate objects, whose articulated arm reminds us of a human arm.

ROBOTS

What is a robot?

The word "robot" was coined by the Czechoslovakian writer, Karel Capek, in 1920. He used it for the main characters of his play who were automatons, similar in all ways to men. The robots could do any kind of job normally done by laborers. The expression comes from the Czech word, "robata," which means forced labor or serfdom. Since Capek used the word, it has entered into all languages to indicate any programmable machine, capable of taking the place of humans in doing manual work of varying degrees of complexity and, often, doing dangerous jobs.

Above you can see a robot playing chess and, on the right, a robot musician. They are robots with vaguely human features, made for demonstration or advertising purposes. The most common type of robot, instead, is like those in the picture on the opposite page, right. They are simple mechanical arms which have very few human characteristics. Here they are being used to make a car in an Italian automobile factory.

Are there many different kinds of robots?

There is a wide range of possibilities, depending upon the type and the complexity of the operation that the robot does. At one extreme, there are very simple robots which can make a limited number of movements, for example, move something from one place to another. At the other extreme, there are quite sophisticated robots, able to make autonomous decisions within certain limits or to use the senses, such as the sense of sight or touch. There are also remote-control robots. What they all have in common, other than "muscular apparatus" is a "brain," which is a computer.

What role does the computer have?

The computer's job is to direct the operation of the mechanical arms and the grasping mechanisms of the robot, giving them the instructions from a man-made program and transcribed, in advance, on magnetic tapes or special perforated boards. The computer "reads" the program and translates the orders into electrical impulses which activate the moving parts of the robot, making it perform the assigned job. The robot, therefore, is nothing but a special case of automation, described in the last two pages, one which simulates the actions of human hands and arms.

Can robots learn?

There are robots that learn, in a certain sense. Rather than writing a program which controls all the movements, the robot's arm is made to perform the desired series of movements. This sequence is recorded by the computer, equipped with a special program, and then followed again and again.

DID YOU KNOW...?

What is a microprocessor?

It is a processing unit of a computer that fits on one chip. A chip is a silicon wafer that has all the necessary electronic circuits on it. In a microprocessor which, when closed in its protective container, is smaller than a box of matches, there is room for millions of elementary electronic circuits.

What are the advantages of microprocessors?

Because they are so small, they allow for the construction of small but powerful computers, like personal computers or small home computers. They also make it possible to insert the equivalent of a computer in other apparati. For example, certain modern household appliances, like the most advanced washing machines, contain microprocessors. This development makes it possible to have more "intelligent" home appliances which can automatically do different functions or regulate themselves in given situations.

Can a washing machine have a memory disk to store the programs?

No, that isn't usually necessary. For more intelligent household appliances, it isn't necessary to incorporate an entire computer. What is basically important is the calculating capacity of the central unit, the microprocessor. If the programs are relatively few and and don't change, they, in their turn, can be memorized on special electronic circuits, called ROM or Read-Only Memory.

Can't the contents of ROM's be erased?

It would be very difficult to erase them, though it might happen due to breakdowns or damage. The possibility of electrical devices breaking down, however, is very much less than that of mechanical devices because they have no moving parts and they don't wear out. The ROM, therefore, reliably saves programs which must not be changed.

Is there also ROM memory in computers?

Yes, in all computers there is some Read-Only Memory. Since a computer doesn't know what to do without a program, from the moment it is turned on, it needs directions, for example, to accept keyboard commands, send information to the terminal and so on. The first program that the computer uses can't be on a disk because even reading information on a disk requires a program! The solution to this problem lies in a ROM within the computer which contains a short program that runs automatically every time you turn on the computer. This short program gives the computer a few basic instructions which serve to activate the keyboard and the terminal and usually to load the operating system from the disk.

What does "load from the disk" mean?

For a program to begin working, it must be in the memory of the computer. To load from the disk means to make the computer read the program on the disk and to put it in memory. This operation is done by giving a command from the keyboard, or by some other means, depending on the computer, to the operating system.

How is the memory made?

The memory of a computer, or Random Access Memory (RAM), is made of electronic circuits like the ROM, but it does not save data in an uneraseable manner. It loses data when the current is turned off or when a command to cancel is given.

What is a modem?

It is a device that can connect a computer to a telephone. It receives electric signals from a computer, transforms them into auditory signals adapted for telephone lines and transmits them. At the destination, another modem receives the auditory signals, transforms them into electric signals and communicates them to another computer. In this way, two computers can have a long distance "conversation" just as we do with a telephone.

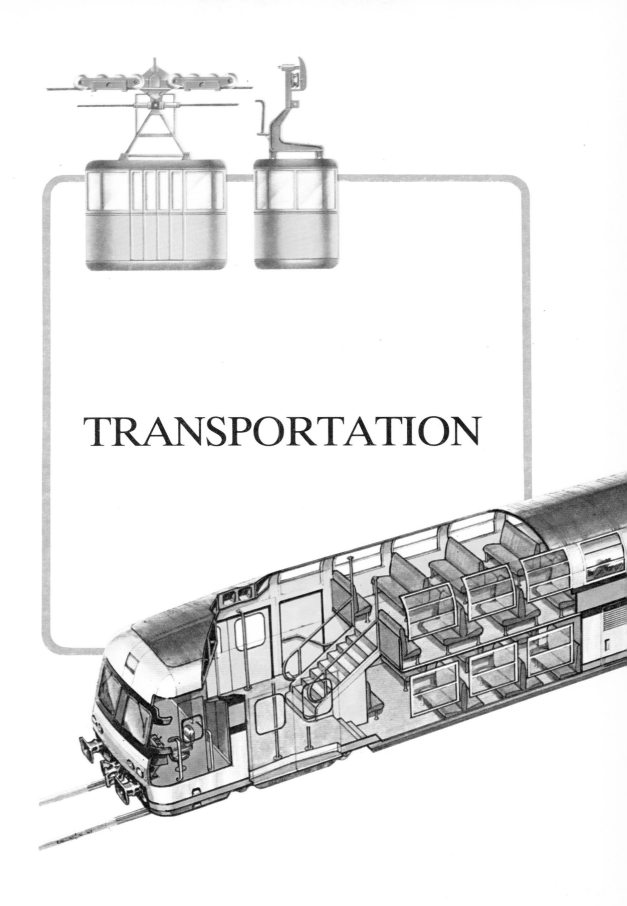

TRANSPORTATION

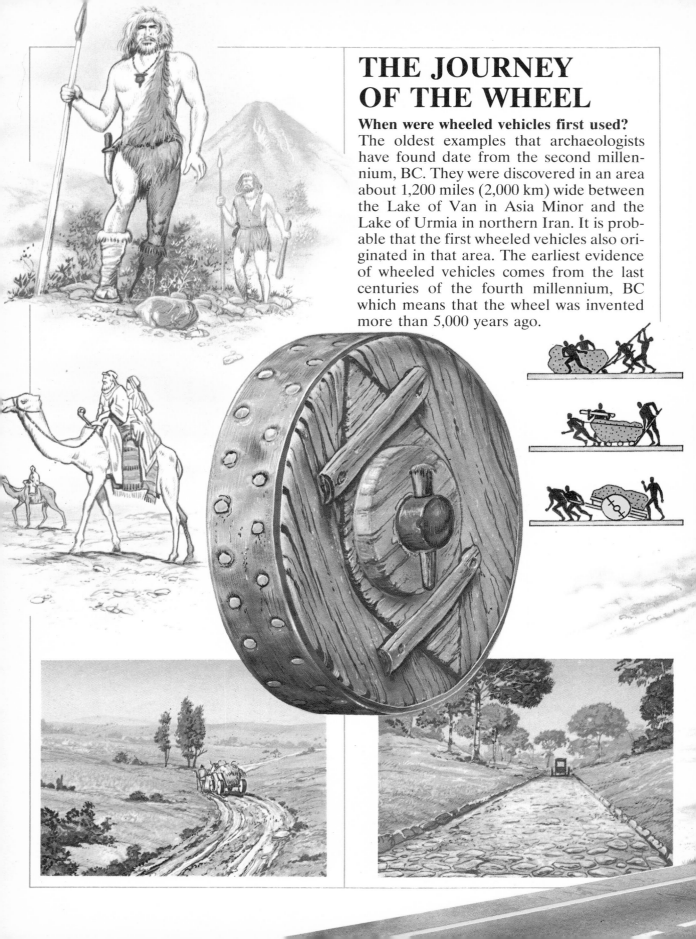

THE JOURNEY OF THE WHEEL

When were wheeled vehicles first used?
The oldest examples that archaeologists have found date from the second millennium, BC. They were discovered in an area about 1,200 miles (2,000 km) wide between the Lake of Van in Asia Minor and the Lake of Urmia in northern Iran. It is probable that the first wheeled vehicles also originated in that area. The earliest evidence of wheeled vehicles comes from the last centuries of the fourth millennium, BC which means that the wheel was invented more than 5,000 years ago.

Before the wheel, how were people and things moved?

They were moved by muscle power, first by the arm and leg muscles of men and later by using animals, such as oxen or camels, which could pull simple sleds.

Did the first wheels have spokes?

The oldest wheels that we know of were made either from a single large board or with many axles joined together (see opposite page). Spoked wheels, along with faster carts and domesticated horses from the Steppes, were among the first solutions to the military needs of the Near East in the first half of the second millennium, BC.

Was the development of roads important for wheeled transport?

There is a close relationship between the construction of roads and the types of wheels that can travel on them. Spoked wooden wheels were suitable for country roads; carriages with metal-reinforced wheels ran well on paved roads. Stagecoaches were good for the beaten tracks used by the American pioneers. But it would be unthinkable today to make a bus like the one shown at the bottom of the page if it weren't for the network of asphalt roads that cover the globe. Development in road building runs parallel to progress in transportation.

What is a tire?

It is what covers the wheel in most modern vehicles from bicycles to automobiles. It is tubular in shape and very flexible. Made of natural or synthetic rubber, it is inflated with pressurized air.

What is a radial tire?

It is a tire which, in the carcass or internal part, contains nylon or other textile fibers, sometimes one on top of the other. This makes the tire more compact, more resistant and better able to adhere to the road.

STEAM MACHINES

How does a steam machine work?

Think of a covered pot in which water is boiling. The boiling produces steam which possesses enough energy to lift the lid of the pot. Steam machines are based on this principle: by putting a lot of water in a large boiler and heating it until it boils, you can produce a large amount of steam. If you then let the steam escape through a tube, it contains a great deal of energy and is capable of doing work, for example, driving a piston back and forth in a cylinder. If the cylinder is connected to a system of wheels, its movement can make the wheels turn. This is how the old steam locomotives worked.

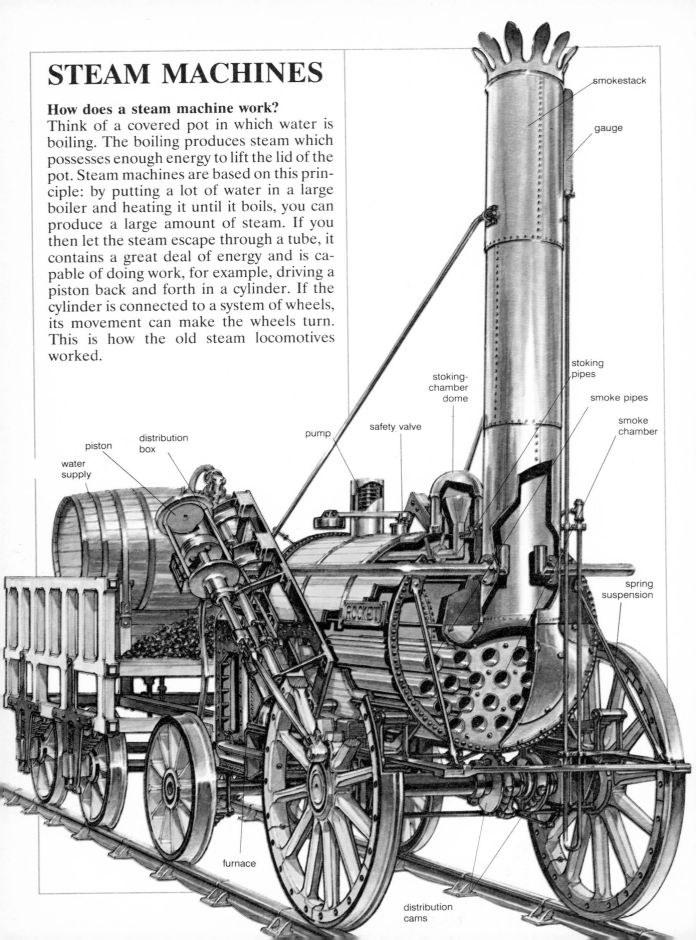

smokestack

gauge

stoking pipes

stoking-chamber dome

smoke pipes

smoke chamber

pump

safety valve

spring suspension

piston

distribution box

water supply

furnace

distribution cams

Is the steam engine a thermal engine?

Yes, it is. We use the expression "thermal engine" for all engines whose operation is based on the transformation of thermal energy, or heat, independently of how it is produced. The more heat that the engine can transform, the greater the mechanical work it can do. The work is usually produced by a gaseous substance, like steam, within the engine. Such substances are called active fluids and heat carriers. That means that heat is introduced into the engine at given points in the cycle and given off at other points.

Are there different kinds of thermal engines?

There are two basic types of thermal engines: internal combustion and external combustion. The difference lies in whether the fuel that supplies the heat is burned within or apart from the active fluid.

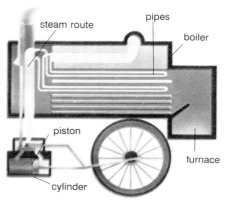

Are steam engines external combustion engines?

Yes, because the fuel (for example, coal for locomotives) is not in contact with the water which is heated.

How is a steam locomotive made?

A true steam locomotive is made up of an engine on which sits the boiler coupled to a tender, which transports stoking fuel and water.

How is the engine of a locomotive made?

The engine includes the burner which, by burning fuel, produces steam under pressure. When the steam expands, it supplies the energy needed to move the train. It also includes the moving apparatus which transforms the steam energy into the rotary movement of the wheels. The wheels, burner and other operating parts are connected to a sturdy frame.

What was the Rocket?

The Rocket was the name of the locomotive built by George Stephenson which, in 1829, began running between Liverpool and Manchester. The Rocket is drawn on the opposite page. On the left, you can see a diagram of its operation. The speed of Stephenson's locomotive was already noteworthy: the Rocket could go about twenty-seven mph (47 km/h).

EARLY AUTOMOBILES

When were automobiles first invented?
The first attempts to substitute horse-drawn vehicles with mechanical systems were made in the eighteenth century, and steam engines were the first type tried. The first internal combustion engines, from which modern car engines are derived, appeared in the last decades of the nineteenth century.

When was the first automobile with an internal combustion engine built?
It was built in 1886 by Karl Benz in Germany and moved at about 10 mph (16 km/h).

What is this vehicle on the right?
This is the "fire wagon" built in 1769 by the French engineer, Joseph Nicolas Cugnot. It is the first example of a steam engine "automobile," capable of hauling heavy artillery pieces during military campaigns.

How was it made?
It had three wooden wheels encircled by iron and a heavy wooden frame which supported a large pear-shaped copper burner. The burner had two walls. The fire was lit in the lower part between the two walls; the internal part contained the water. The steam produced by the burner was introduced into two large cylinders, which had a combined capacity of fifty liters.

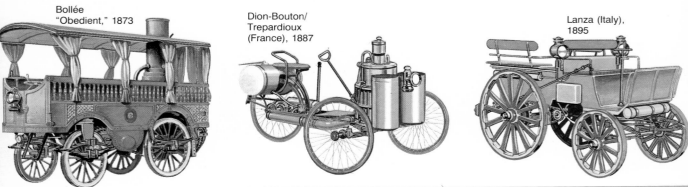

Bollée
"Obedient," 1873

Dion-Bouton/
Trepardioux
(France), 1887

Lanza (Italy),
1895

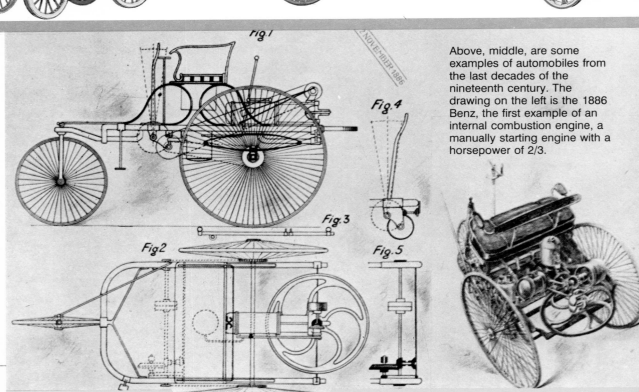

Above, middle, are some examples of automobiles from the last decades of the nineteenth century. The drawing on the left is the 1886 Benz, the first example of an internal combustion engine, a manually starting engine with a horsepower of 2/3.

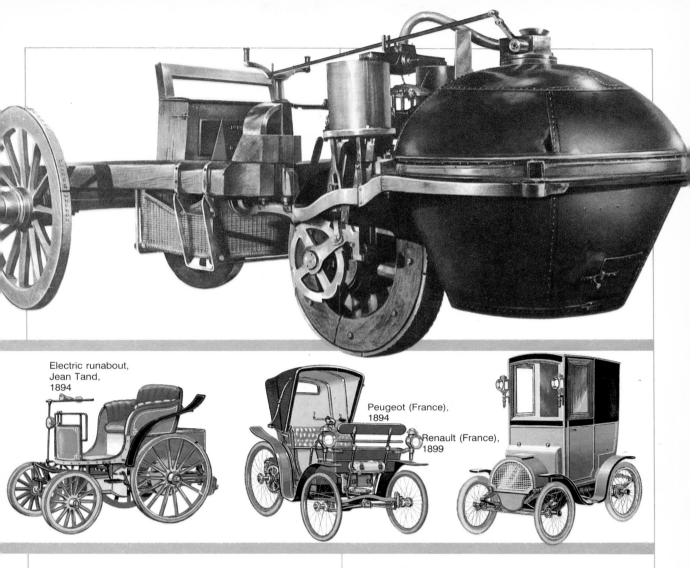

Electric runabout,
Jean Tand,
1894

Peugeot (France),
1894

Renault (France),
1899

What speed did Cugnot's vehicle reach?

It managed to go at a velocity of 1.8 to 5.4 mph (3-6 km/h), with four people on board and pulling a load of four or five tons. It had the problem of limited autonomy because it was not possible to refill the water in the boiler easily. After fifteen minutes of marching, the vehicle had to stop and wait another fifteen minutes for the engine to cool down enough to put in more water.

Which was the first popular automobile?

It was the Model-T produced in 1908 by Henry Ford, a car with an internal combustion engine built with modern standards and mechanical solutions that are still valid today.

How many Model-T cars were made?

The Model-T remained in production until 1927. Eighteen million of these cars were produced in that time. Because of the Model-T, the automobile became accessible to many people. It was designed to be economical, and even the workers who made this car could afford to buy it.

How much did it cost?

Its price continued to go down. In 1908, it cost $850.00, but in 1924, only $290.00. The constant increase in production and decrease in cost was due to a particular method of working which Henry Ford adopted and which is still used today in all kinds of industries: assembly-line production.

MODERN AUTOMOBILES

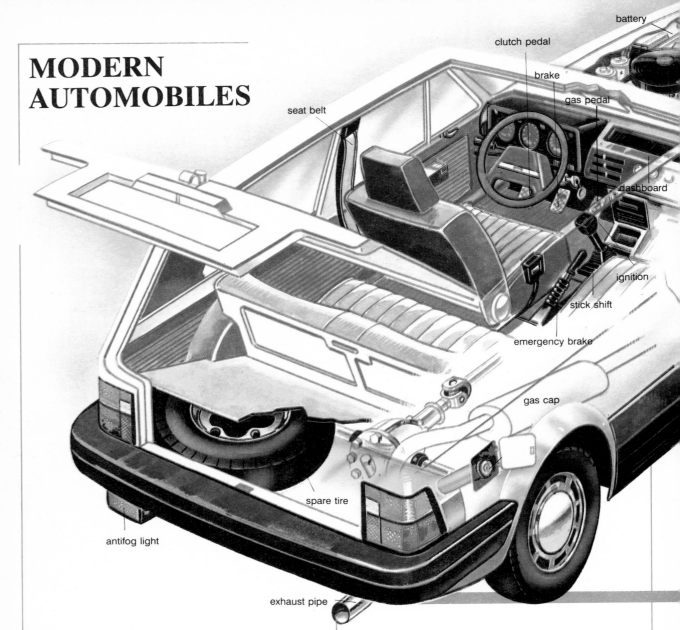

battery

clutch pedal

brake

gas pedal

seat belt

dashboard

ignition

stick shift

emergency brake

gas cap

spare tire

antifog light

exhaust pipe

How does the gasoline engine of a modern car work?

Modern cars have internal combustion engines. A mixture of gas and air is introduced into a chamber or cylinder, the upper part of which can move in place. This part, called the piston, moves up and down within the cylinder. When the piston moves down, it compresses the fuel in the cylinder. At this point, an electric spark from a spark plug makes the gas explode. This explosion pushes the piston up. This cyclical procedure continues as long as the automobile is running.

What is a crankshaft?

The piston is connected to a crankshaft, a metal rod, whose shape makes it possible to transform the alternating movement of the piston into rotary motion, turning another rod, the drive shaft.

What is the drive shaft connected to?

Usually gears connect the drive shaft to the rear axle and sometimes to the front axle. These axles are the metal bars that connect the back wheels (rear axle) and the front wheels (front axle). The movement of the drive shaft turns the axle which makes the wheels turn.

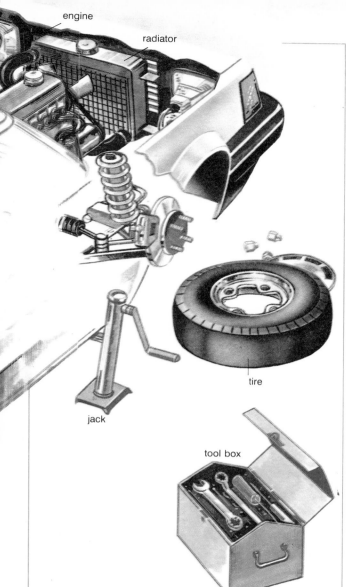

engine

radiator

tire

jack

tool box

Are there automobiles that don't run on gasoline?

Yes, there are automobiles with diesel engines whose operation is a little different from those with gasoline engines (eight-cycle engines). There are also electric vehicles which, at this point, are still in the experimental stages. (Two examples are shown below.) These cars evoke a lot of interest because it is hoped that they may solve the problems of energy consumption and air pollution, since they do not burn fossil fuels. In the first example below, the electric energy comes from panels of photovoltaic cells, which use solar energy. These panels almost completely cover the body of the car. The second example, a smaller car, runs on batteries.

What does the stick shift do?

The stick shift transmits the engine power to the axles. It connects the crankshaft to the drive shaft and makes it possible to change the relationship of the gears of these two shafts. The change from one gear to another involves a variation in power that the engine must produce to make one complete turn of the wheels. It, therefore, changes the amount of energy needed to go a given distance.

What is the brake?

It is the piece of equipment that connects or disconnects the crankshaft from the transmission shaft.

RACE CARS

When was the first automobile race?
The first road race for automobiles was run in 1895 and was won by a Panhard car. The winning vehicle managed to do the course at an average speed of fifteen mph (25 km/h).

Are the tires important?
Since a race car can reach very high speeds, the tires must adhere to the road very well. Special tires are developed for various types of road surfaces and different weather conditions. These tires quickly wear out. During a race, the tires are often replaced (see the photograph below).

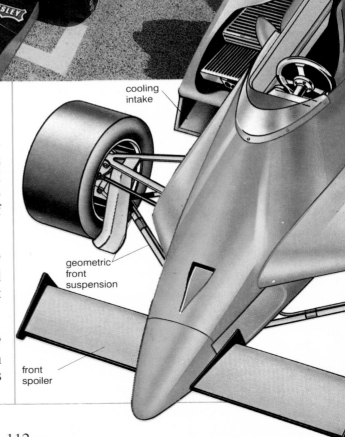

turbocompressor

cooling intake

geometric front suspension

front spoiler

How many cylinders does an automobile have?
An ordinary car, like the ones we use everyday, has from four to eight cylinders. Four is the usual number for European cars and eight for the large American cars. Each cylinder has a piston which does an average of 5,000-6,000 rpms.

How many cylinders are found in race cars?
They usually have twelve cylinders and their engines run at higher speeds, about 11,000-14,000 rpms.

Are race cars different in any other ways?
Yes, they are quite different, even though the basic operating principle is the same as that of an ordinary car.

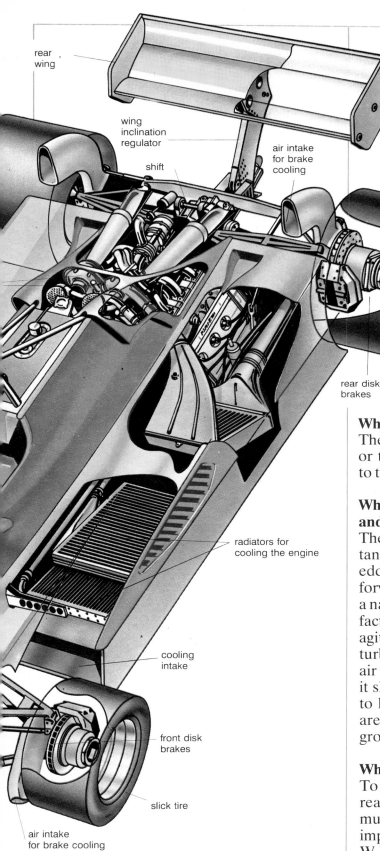

rear wing

wing inclination regulator

shift

air intake for brake cooling

rear disk brakes

radiators for cooling the engine

cooling intake

front disk brakes

slick tire

air intake for brake cooling

The body of a modern race car weighs one-third or one-fourth as much as an ordinary car. The major part of the weight is concentrated in the engine, the transmission and the wheels. Each of the points marked here has required a lot of study, to make use of new materials, to reduce aerodynamic drag or to get more sturdiness and rigidity.

What is the body of a race car made of?
The materials typically used are fiberglass or titanium which are very light compared to the steel generally used for ordinary cars.

Why are race cars so low to the ground and shaped in that way?
These things help to overcome air resistance. Air, when pushed aside, creates little eddies which slow down and unbalance the forward thrust of the car. The advantage of a narrow- nosed, steamlined body lies in the fact that, because of its reduced size, it agitates a minimal amount of air, reducing turbulence as much as possible. Even the air under the car is turbulent and can make it skid. For this reason, race cars are made to hug the ground, and the recent models are only about eight inches (20 cm) off the ground.

What is the Kamm Tail?
To reduce air turbulence under the car, the rear end of the racer is usually shortened so much so that it seems to be cut off. This improvement is named after its inventor, W.I.E. Kamm.

RALLY CARS

What is a rally?

It's an automobile race which is run either on special closed tracks or on roads open to the public. In the second case, the competitors must follow traffic laws. The length of the race varies from hundreds to thousands of miles.

How are rally cars made?

They are ordinary, mass-produced cars, somewhat modified to increase their sturdiness. (Rally races are often run over very difficult terrain.)

Are rallies only racing competitions?

Like all automotive races, they are competitions. But their importance goes beyond that as they often serve as difficult and effective "proving grounds" for future technological developments.

What is front-wheel drive?

Front-wheel drive is when the wheels connected to the engine, those that it moves, are at the front. Cars built like this are pulled forward rather than being pushed as occurs with rear-wheel drive. Front-wheel drive vehicles hold the road better.

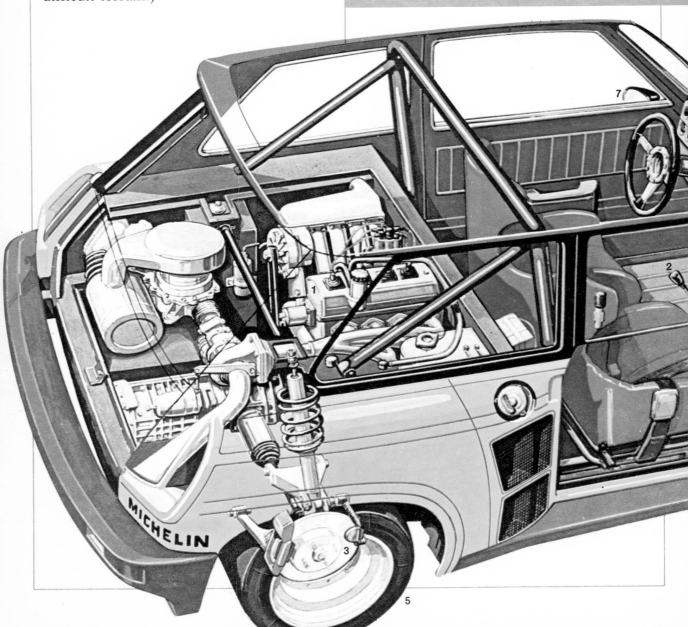

5

Section of a typical rally car. The engine (1) is supercharged by a turbo-compressor. The displacement is 1,397 cc the weight is about 1980 pounds (900 kg), and the speed exceeds 120 mph (200 km/h). The shift (2) has five gears, and there are four disk brakes (3). The engine occupies a central position. There are only two seats and the spare tire

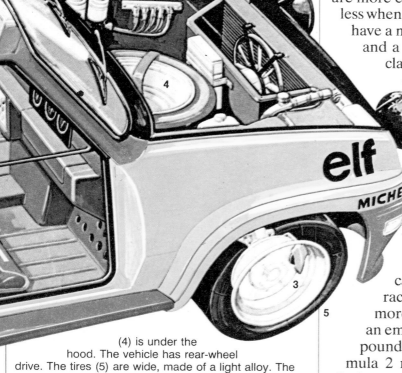

(4) is under the hood. The vehicle has rear-wheel drive. The tires (5) are wide, made of a light alloy. The two gas tanks, with a capacity of about, 395 gallons, are under the seats. The dashboard (6) gives complete instrument readings. The rear-view mirror (7) is adjusted electrically. The fenders are made of polyester.

What is a turbo-compressor?

It is a device which compresses the air for the air-gasoline mixture before it is introduced into the cylinder and increases engine performance.

What are disk brakes?

They are replacing drum brakes, since they are more effective and sturdier and heat up less when used for a long time. Disk brakes have a metallic disk applied to the wheel and a caliper which, when you brake, clamps the brake pads against the disk, producing strong friction that stops the car.

Are rally cars and Formula 1 race cars different?

Rally cars are ordinary vehicles with few modifications; Formula 1 cars are special vehicles. They are not mass-produced and must respect weight and measurement standards fixed by international regulations. The cars that participate in Formula 1 races have a displacement of not more than 3,000 cc and a weight, with an empty gas tank, of not less than 1278 pounds (585 kg). There are also Formula 2 races, for cars with a maximum displacement of not more than 2,000 cc and a weight of 1045 pounds (475 kg), and Formula 3 cars, with a displacement of 2,000 cc and engines from four-cylinder road cars.

CROSS-COUNTRY VEHICLES

What are cross-country vehicles?
They are vehicles that can travel over terrain that would be impossible for conventional cars to handle: in meadows, on mountain slopes, through rocky streams, on desert sands and over difficult mule tracks. To perform in this way, they have some characteristics which distinguish them from ordinary road cars.

Is the engine of a cross-country vehicle different from an automobile engine?
The engines of both vehicles work in the same way, but the cross-country engine is sturdier and more powerful. It is, moreover, often a diesel engine.

What type of drive do these vehicles use?
They usually have four-wheel drive, which means that the engine can be connected to all four, or sometimes six, wheels and not only to two, as is the case with conventional vehicles. All the wheels, therefore, are set in motion and drive the vehicle. A vehicle equipped with four-wheel drive has an advantage over conventional cars on soft or slippery ground because its wheels grip well.

Is the cross-country vehicle in the illustration typical?
This automobile, a Suzuki SJ 410, is an all-purpose cross-country vehicle. With its refined finishing and comfortable interior, it can also be used for city driving.

Do the wheels of a cross-country vehicle also have special characteristics?

Yes, cross-country wheels have a larger diameter than wheels of normal cars. They can, therefore, better absorb the effects of holes in the ground. The tires are broader to adhere better to the road and sink less in soft terrain. They also have a deeper tread to grip the road better and to avoid skidding.

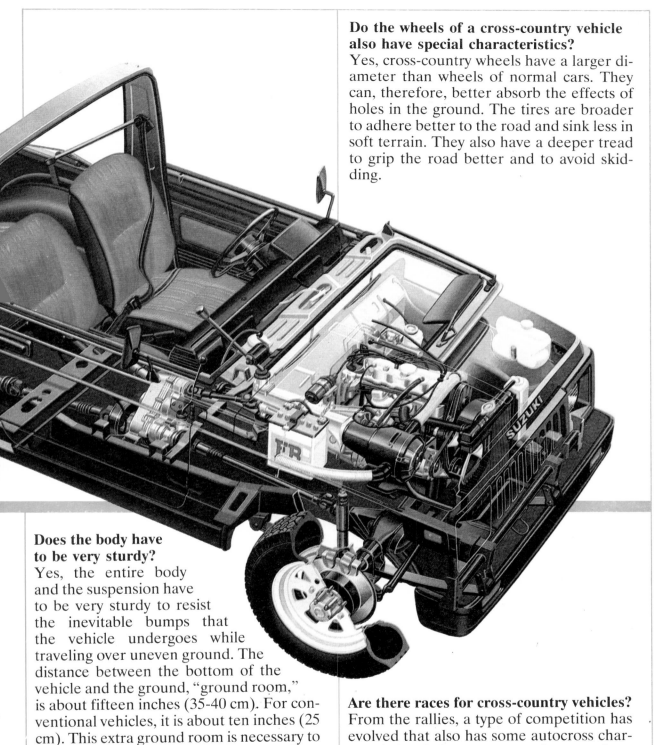

Does the body have to be very sturdy?

Yes, the entire body and the suspension have to be very sturdy to resist the inevitable bumps that the vehicle undergoes while traveling over uneven ground. The distance between the bottom of the vehicle and the ground, "ground room," is about fifteen inches (35-40 cm). For conventional vehicles, it is about ten inches (25 cm). This extra ground room is necessary to keep the lower parts of the vehicle from bumping against the ground and to allow the vehicle to ford rivers and streams as deep as twenty-seven inches (70 cm). A normal car would be in difficulty in water twelve or more inches deep.

Are there races for cross-country vehicles?

From the rallies, a type of competition has evolved that also has some autocross characteristics. It is a race over desert and savannah in hostile environments. Examples are the Paris-Dakar, which crosses the Sahara Desert, and the Camel Trophy, for four-wheel drive vehicles. (The opposite page shows a competing vehicle.)

AUTOMOBILE PRODUCTION

How many cars are built in one day?
The large automobile factories can produce more than one thousand cars per day.

How is a new automobile created?
The creation of a new model begins with the planning which takes many elements into account: the public demand, the types of cars produced by competitors, the characteristics which the car should have to sell well and so on. After various development phases, tests and modifications, a prototype is built.

How is a prototype tested?
It is tested under all possible conditions, on the road and cross country. Only after it has successfully passed all these tests is it ready for mass production.

What is an assembly line?
It is the sequence of phases in the assembly of mass-produced automobiles. Nowadays a large part of the procedure is automatic. The elements and pieces to be assembled are carried on conveyor belts or special rollers to the various stations. At each station, only certain operations are done, manually or with automatic tools. The process goes from the cutting of the panels in large presses to the finishing of the vehicle in the final stage. (A simplified diagram of this process is shown below.)

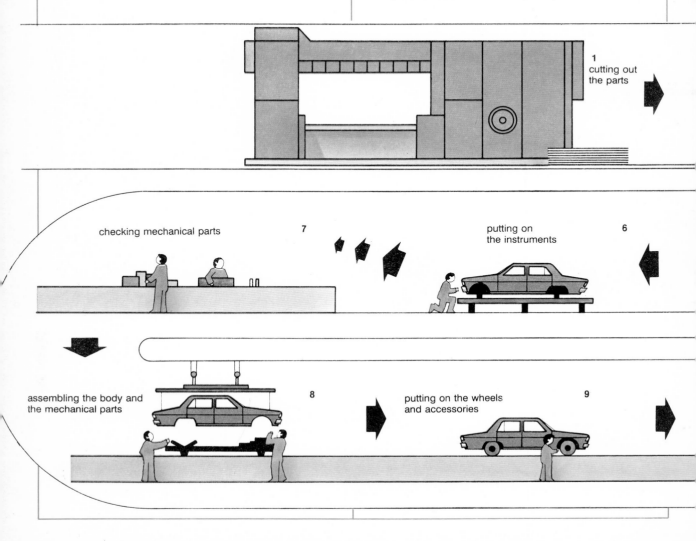

1 cutting out the parts

7 checking mechanical parts

6 putting on the instruments

8 assembling the body and the mechanical parts

9 putting on the wheels and accessories

Are computers used in the planning?

They often are. There are special programs for the technical design and planning of cars. The computer stores the data that the technicians have given it, and on this basis does very fast calculations, makes diagrams and, using sequences of images that resemble animated drawings, even carries out tests which, later on, will have to be done on a real vehicle.

Are there also robots on the assembly line?

Yes, there are. You can see one in the photograph on the left. This one, which is essentially a mechanical arm, is screwing in a part. Others may be used for welding or painting. The automobile industry was one of the first in which automation, and in particular the use of robots, found a wide field of application.

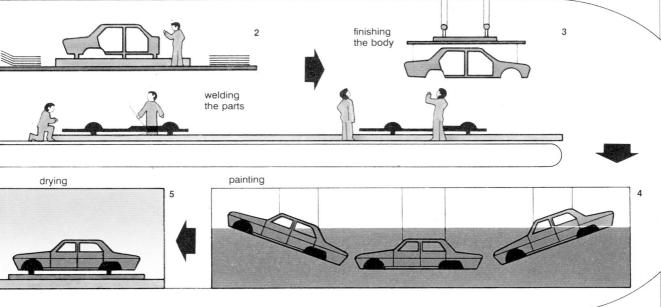

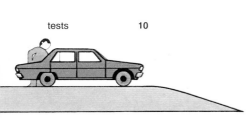

The principal phases in an assembly line: (1) large presses for cutting the steel parts that go into making the body. (2) Separate welding for the parts that make up the lower and the upper sections. (3) Final welding of the various elements to complete the body. (4) Painting by immersion. This is a complex phase that includes an anti-rust treatment, a protective paint and the use of special processes for applying the last coat of paint. (5) Passing through a tunnel for drying and baking of the last coat of paint. (6) Beginning to put on the controls, internal instruments, electrical system and so on. (7) Being sent, by overhead conveyor, to the production department, and the testing of the engine group and other mechanical parts. (8) Putting together the body and the mechanical parts. (9) Finishing the construction with wheels, interior upholstery and so on. (10) The final tests before starting the road tests.

GARAGES

What is a garage?

It is a place where we keep a car and, if necessary, where we also repair it. In English, the same word is used to mean a private garage, a public garage, where you usually pay to park your car and a repair shop, where you can get help to fix your car. The first garages were shelters for rebuilt car bodies.

Do private garages have special characteristics?

Like all buildings, private garages must follow detailed building codes, established by law. Since automobiles emit gas and inflammable oils, fire-control laws specify that the walls and ceilings of garages must be treated so as to reduce to a minimum the risk of a fire spreading to nearby houses.

motor

electromechanical bridge

tools

tire-changing equipment

battery charger

alternator tester

pit tester

anti-skid floor

fireproof metal tool cases

pit (not in use)

jack

battery being charged

120

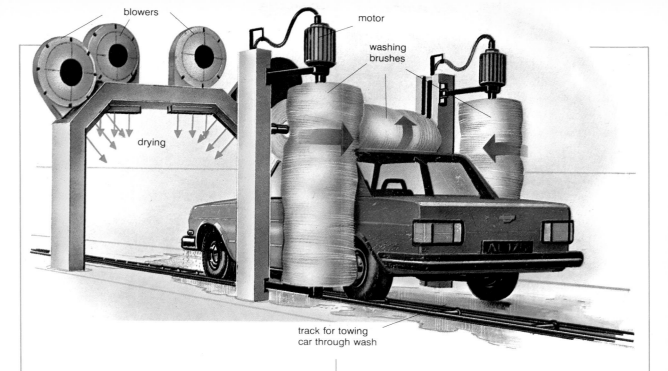

blowers

motor

washing brushes

drying

track for towing car through wash

What are the lifts in garages used for?

Hydraulic and/or electric lifts are used to jack up vehicles that need repair or maintenance. Most garages have hydraulic lifts which raise cars by means of a pressurized oil pump system and keep them raised, securely attached to either a center post or two side posts. The lifting is done by placing braces on the strong parts of the frame and then using an electric motor which activates a chain or hydraulic cylinders.

What other equipment is used in garages?

You can see several pieces of equipment in the drawing on the opposite page. A vehicle can easily be lifted with a movable jack best used in restricted spaces. Near the hydraulic jack, there are many useful tools: portable lights, air compressors for jet cleaning and guns for spraying oil and greases. In the background, you see equipment for maintaining the electric parts of the vehicle (for testing batteries, generators, alternators, rectifiers and to recharge connectors-disconnectors) and, in the middle, a work area with the most commonly-used tools.

How do automatic car washes work?

A car is placed on parallel tracks which move it forward at a speed that is regulated by the speed of the washing equipment. (See the example above). Nozzles spray water and detergent onto the car, and a series of rotating brushes cleans the body with an energetic rubbing action. Of course, the action is not strong enough to damage the car. After the car passes through the brushes, it is rinsed and the remaining detergent is taken off. In the last stage, strong blowers dry the car.

Are motorized tools used in a garage?

While mechanics use several hand tools, such as screwdrivers, sockets and wrenches, they also employ a variety of motorized ones. The pneumatic, or air-powered, ratchet quickly twists off nuts and bolts while the impact wrench removes larger parts, such as wheels and axles. These pneumatic tools use compressed air rather than spark-producing electricity in the motors, which is a significant safety factor in the inflammable area of the garage. Many important automobile functions are now controlled by refined computer systems.

MOTORCYCLES

When was the first motorcycle made?
It was made in 1869 in France and had a steam engine.

When was the first motorcycle to contain a gasoline engine built?
It was built about ten years later, in 1879.

Are there many different kinds of motorcycles?
There are road, sports and cross-country bikes. Their displacement varies from fifty to more than 1,000 cc, equivalent to certain cars. Bikes may have two- or four-stroke engines and from one to six cylinders. Their horsepower ranges from five or six to more than one hundred for the maxi-bikes.

What does two- or four-stroke mean?
This expression refers to the way in which the operating cycle of the cylinder is conceived. In four-stroke engines, two turns of the shaft are needed to complete a cycle, that is intake, compression, expansion after sparking and output. A two-stroke cycle requires only one turn of the shaft.

What is the displacement?
It is the volume covered by the course of the piston, as it passes from the lowest to the highest point of the cylinder. When the cylinders are the same, the displacement value, which is usually expressed in cubic centimeters, cc, or liters, l, is found by multiplying the volume displaced by one piston by the number of cylinders.

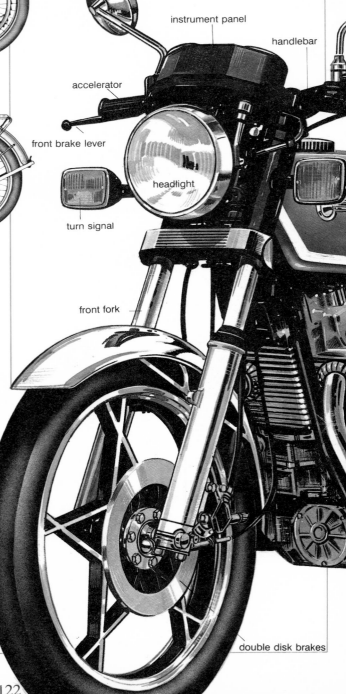

instrument panel

handlebar

accelerator

front brake lever

headlight

turn signal

front fork

double disk brakes

What are the motorcycles that are pictured on these two pages?

They are examples of important motorcycles from different eras. On the opposite page, at the top, is a Norton "Oldmiracle", an English bike from 1912, with a 490 cc engine. Below it is a Harley-Davidson 11, a three-speed American bike from 1915, with a two-cylinder 989 cc engine, 11 hp. On this page, at the top, is a Garelli 350 three-speed racing bike, made in Italy in 1924, with a 348 cc two-cylinder engine, 20 hp, and a weight of only 46 pounds (100 kg). In the middle, a Triumph Grand Prix, a four-speed English bike from 1947, with a two cylinder 500 cc engine. At the bottom, a BMW R90S, a German bike made in 1975, with a two-cylinder 898 cc engine, 67 hp, weight, 98 pounds (215 kg), and maximum velocity of more than 120 mph (200 km). The large drawing between these pages is a Kawasaki GP 1100, a typical example of a refined maxibike.

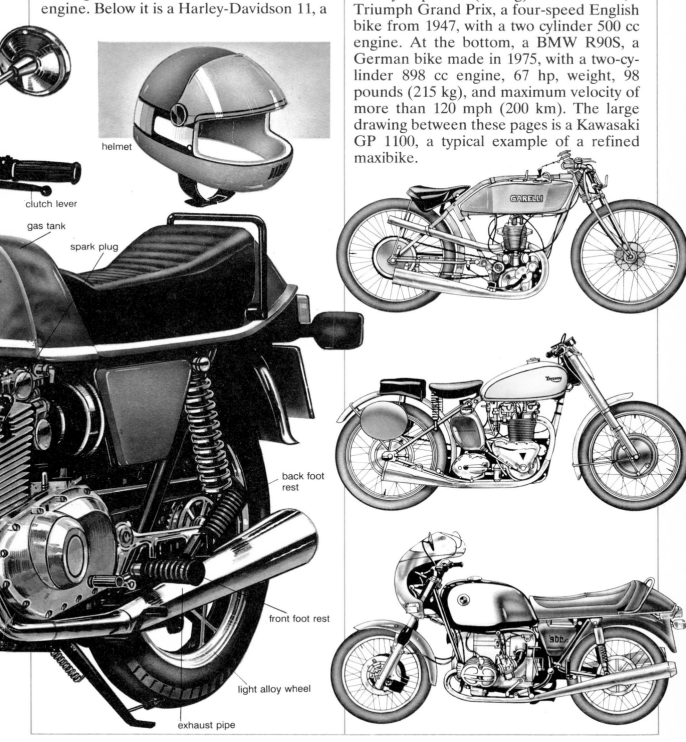

helmet

clutch lever

gas tank

spark plug

back foot rest

front foot rest

light alloy wheel

exhaust pipe

BICYCLES

How is a bicycle frame made?

It is made up of three triangles, two of which are parallel and hold the back wheel in position. The third forms the body between the two wheels. The tubes of the back triangles hold the saddle in place. The tubes attached to the back wheel are called "forks" because they fork slightly at the ends to provide flexibility. The "cross" frame bicycle, invented in 1962 by A. Moulton, has a single-tube structure from which the supports for the saddle and back go off at right angles.

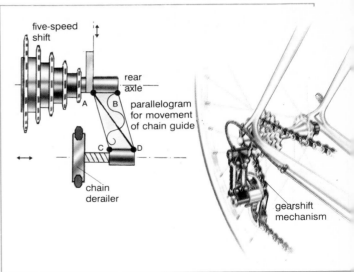

five-speed shift

rear axle

parallelogram for movement of chain guide

chain derailer

A B C D

gearshift mechanism

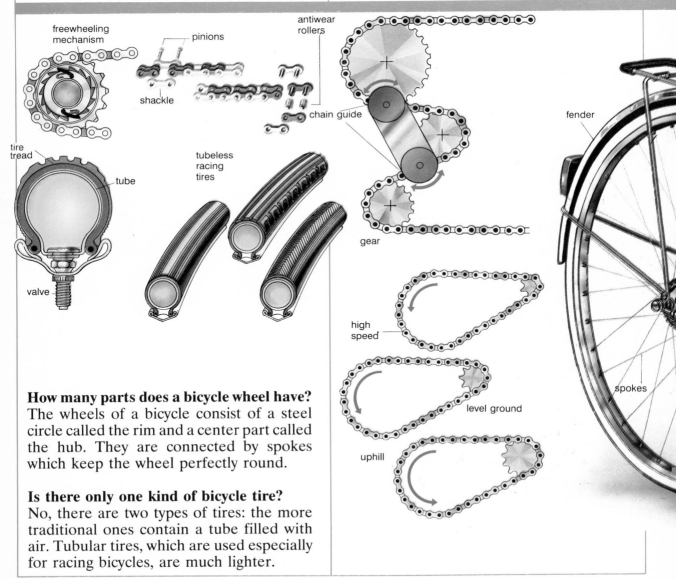

freewheeling mechanism

pinions

shackle

antiwear rollers

chain guide

gear

tire tread

tube

tubeless racing tires

valve

fender

high speed

level ground

uphill

spokes

How many parts does a bicycle wheel have?

The wheels of a bicycle consist of a steel circle called the rim and a center part called the hub. They are connected by spokes which keep the wheel perfectly round.

Is there only one kind of bicycle tire?

No, there are two types of tires: the more traditional ones contain a tube filled with air. Tubular tires, which are used especially for racing bicycles, are much lighter.

What is the gearshift for?

It allows the rider to change the speed of transmission, that is, the number of times the wheel rotates with each rotation of the pedals. Shifting gives cyclists the possibility of rationing their efforts according to the characteristics of the terrain which may be, for example, flat or hilly.

How is the gearshift made?

In the drawing on the left, you see the rear wheel axle of a five-speed bicycle. The chain moves on the central derailer of the rear hub and downward to the return wheel. This movement causes the chain to move onto a different gear. To the left of this drawing, you see the sprockets and the parallelogram which allows for the movement of the chain guide parallel to itself. Below you see the rotary mechanism which prevents the chain from running onto the hub when the gear is being changed. Below you see the chain on the three sprockets of the rear hub.

saddle

tool kit

hand brake

handlebars

bell

generator

light

pump

hub

pedal

chain

tires

TRUCKS

What is a truck?
Trucks are vehicles for road transport of goods or for other industrial activity. Since they have internal combustion or diesel engines, they are similar to cars. They are, however, larger and look different. A truck body is normally made up of welded longitudinal pieces. Only in vans or smaller, closed trucks does the body structure bear the weight.

What is a trailer truck?
It is a truck that pulls an independent trailer.

What is a semi-trailer?
A semi-trailer has an engine unit, similar to that of an ordinary truck, which does not, however, transport goods. A semi-trailer is attached or hooked up to this unit. This solution is very useful because the engine unit, detached from the semi-trailer, can make other trips while the trailer is being loaded or unloaded.

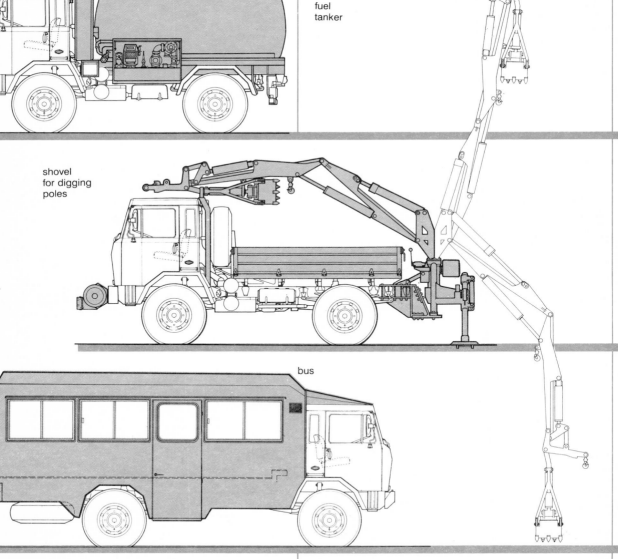

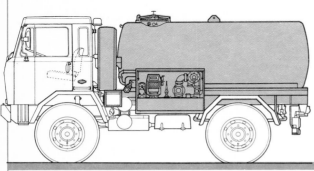

fuel tanker

shovel for digging poles

bus

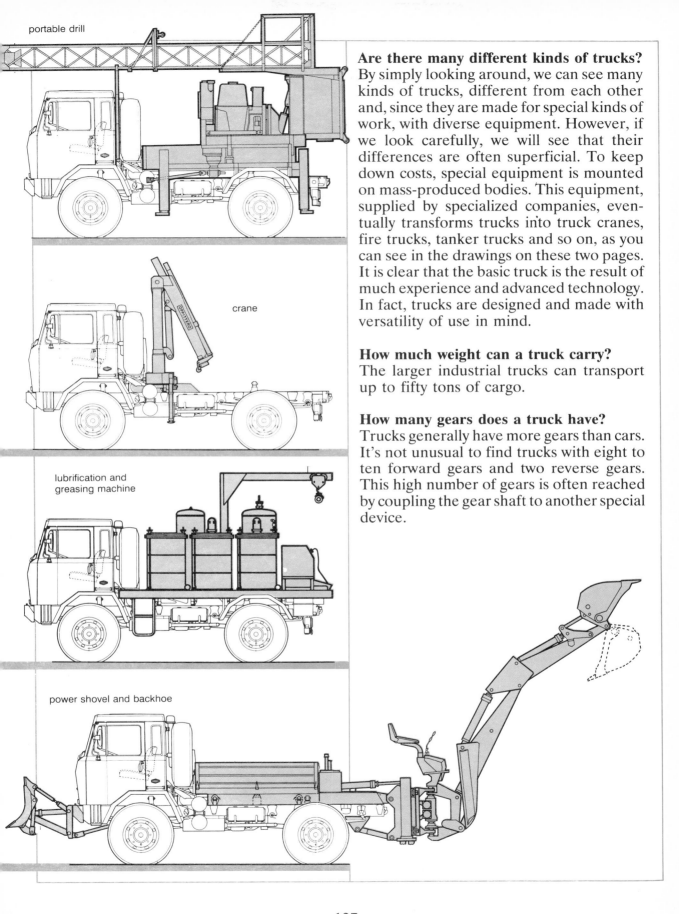

portable drill

crane

lubrification and
greasing machine

power shovel and backhoe

Are there many different kinds of trucks?

By simply looking around, we can see many kinds of trucks, different from each other and, since they are made for special kinds of work, with diverse equipment. However, if we look carefully, we will see that their differences are often superficial. To keep down costs, special equipment is mounted on mass-produced bodies. This equipment, supplied by specialized companies, eventually transforms trucks into truck cranes, fire trucks, tanker trucks and so on, as you can see in the drawings on these two pages. It is clear that the basic truck is the result of much experience and advanced technology. In fact, trucks are designed and made with versatility of use in mind.

How much weight can a truck carry?

The larger industrial trucks can transport up to fifty tons of cargo.

How many gears does a truck have?

Trucks generally have more gears than cars. It's not unusual to find trucks with eight to ten forward gears and two reverse gears. This high number of gears is often reached by coupling the gear shaft to another special device.

127

BULLDOZERS

What is a bulldozer?
It's a very solid machine, weighing from six to forty tons, which is used for digging, spreading, grading and transporting earth in construction projects from homes to highways.

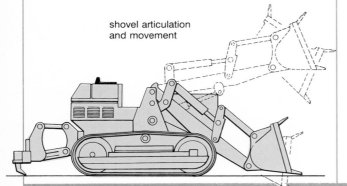

shovel articulation
and movement

What kinds of work does a bulldozer do?
It is useful for digging shallow ditches, transporting materials short distances, spreading soil unloaded by a dump truck, cleaning and leveling around loading equipment and doing rough work such as grading or removing trees, stumps and boulders.

On the left is the movement of the articulated arms for raising the shovel. On the right are three possible blade movements. For side movement use, the angledozer, for transversal movement, the tiltdozer and for up and down movement, the tipdozer.

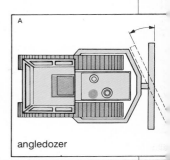

angledozer

How is a bulldozer made?
A bulldozer consists of a tractor – either a four-wheel drive vehicle or a tanklike or "caterpillar" type on continuous metal treads – with a heavy, broad steel blade or shovel mounted in front.

How is the shovel attached?
The shovel is attached to the front of the tractor by means of two arms, the push or lift arm on the bottom and the dump arm on top, which the operator regulates hydraulically with levers. The shovel is raised or lowered according to how the bulldozer is being used. When the shovel is moved below the surface of the earth, the bulldozer can dig. To transport earth, the operator raises the shovel above the level of the ground.

How long have bulldozers been in use?
They have been used since 1945, after World War II.

What horsepower do bulldozer engines have?
The horsepower of a bulldozer engine can vary from fifty to seven hundred, depending on the size of the machine and the type of work that it does.

How is a bulldozer used in highway construction?
Bulldozers are especially useful in highway construction for clearing trees and moving earth.

What is a ripper?
A ripper is a special device with a series of teeth that is mounted on the rear of the tractor to break up large masses of rock so that the bulldozer can carry them.

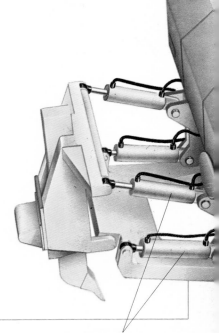

hydraulic pistons
for activating the pick

What are the disadvantages of a bulldozer?

The drawback of a bulldozer is that, in a single pass, it can move earth only as wide as its blade. Nevertheless, it remains a great improvement over a man with a spade.

Is it difficult to maneuver a bulldozer?

It's not very easy to drive a machine like this. A good bulldozer operator knows how to balance the two forces that bear on the machine at once: the forward movement of the tractor and the resistance offered by the soil. If the operator is not skillful, the bulldozer will skid or lose speed, making the engine slow down.

Is it easier to work on a slope?

Yes, pushing earth downhill is easier, and for this reason, many operators try to work downhill. If a pile of earth is too high to be pushed aside in one pass, the operator will drive back and forth until it is moved. The procedure is not as simple as it sounds; it can even be dangerous.

Why is it dangerous?

Since the blade is relatively narrow, high piles of earth can form on either side of the bulldozer and then collapse onto it. To avoid that, the operator must turn into the sides from time to time and push the earth away from the machine.

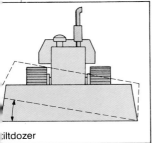

tiltdozer

tipdozer

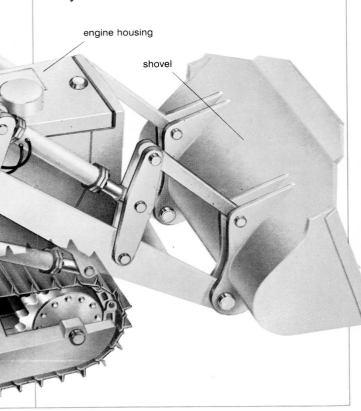

steering lever

shovel and ripper controls

exhaust

muffler

engine housing

shovel

SPECIAL KINDS OF TRANSPORT

Which vehicles are used for special kinds of transport?

There are many situations in which a car or truck is of no use: traveling cross-country, snow-plowing or carrying very bulky or heavy loads. Special vehicles with specific characteristics must be constructed for these situations. Unlike conventional vehicles, which, under normal conditions, can be used for many different reasons, these vehicles usually have one use. For example, special transport vehicles are those that transport exceptionally large or heavy cargo.

What is the vehicle shown on these pages used for?

It's something more than a vehicle. It's a special transport column with two trucks, one at the front and the other at the back, two trailers and the piece being transported, in this case a steam generator weighing 525 tons. Two trucks are needed because of the weight of the merchandise: one pulls and the other pushes. To get an idea of the size of the convoy, look at the man who is in front of the second tractor, below the steam generator.

How does a convoy like this travel on normal roads?

Convoys like this one undoubtedly have considerable difficulty traveling on normal roads with curves and residential areas to negotiate. Special assistance, such as the use of police escorts to keep the roads open, is needed to help them reach their destinations.

What is a truck crane?

It is a crane mounted on a movable base which can run on rubber wheels or on tank-like "caterpillar" treads. Special hydraulic cables or jacks control the angle of the arm which may even reach a length of 150 feet (46 m).

What is a crane used for?

Cranes are used to move heavy loads between points a considerable distance apart and usually at different levels. They are typically used in building construction.

What are self-powered cranes?

Most cranes are fixed equipment and cannot be moved without special equipment. Self-powered cranes are those which can move by themselves because they form part of special vehicles. Truck cranes, for example, are one kind of self-powered crane. Another is the floating crane which is used in ports and at sea. It is usually installed on pontoons or on towing barges. The self-powered cranes have diverse uses, especially in work that is not continuous.

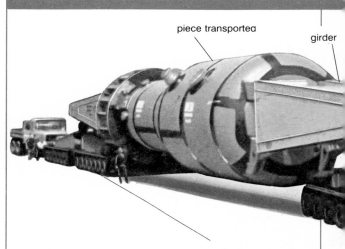

piece transported

girder

The front trailer has 112 tires like the one at the back, but the wheels, all of which can turn, are controlled by the towbar which connects the trailer to the lead truck.

What is a bridge crane?

It is a crane which resembles a metal bridge. It is made up of a lattice beam resting, at both ends, on a guard rail which, in turn, is equipped with trolley runners. By combining the movement of the trolley with that of the beam, it is possible to cover a rectangular surface. These cranes are used in stores because, compared to other cranes, their structure does not interfere with the places where they are used.

Is the vehicle on the left a special kind of vehicle?

Yes, it is a very special vehicle. It is a self-powered gantry crane designed to transport hot-cast metal in a foundry.

What is this type of crane used for?

It is used for moving loads that weigh several tons and for transferring them long distances, hundreds of feet or even miles.

Are these machines very specialized?

Yes, they must be equipped not only to grab and lift hot materials but also to transport these materials far away. They have powerful engines for this purpose. In the foreground of the photograph, you can see the operator control posts for loading, unloading and transferring the materials.

The piece is transported using large articulated girders with supports which rest on the front and back trailers.

The towbar connects the lead truck to the front trailer. Both this one and the one at the back have hydraulic suspension which makes it possible to lower the cargo to the ground when unloading it.

The lead truck or front trailer is helped by the one in the rear. Together they produce 310 hp.

supports

The rear trailer has fourteen axles and 112 tires. All the tires can turn; they are manually controlled by a hydraulic pump run by a diesel motor.

FIRE-FIGHTING EQUIPMENT

What is a fire truck?
It is a truck equipped with pumps and hoses for spraying water or foam onto fires. Fire trucks have a large and powerful pump to which are attached heavy-duty hoses. The pump draws water from a reserve tank on the truck or from nearby fire hydrants. All buildings must have fire hydrants to combat possible fire hazards.

How much water is there in the tank of a fire truck?
There is more than 525 gallons (2000 l).

How long are the hoses?
Each fire truck usually carries several hundred yards of hose.

How much water is pumped through the hoses?
Since fire truck equipment is very powerful, a considerable amount, between 800 and 1,600 gallons (3000-6000 l), of water, can be pumped through the hoses in a minute.

How do you know when to use water or foam to put out a fire?
Water is the best way to extinguish burning solids like wood, cloth, paper and plastic because it can effectively cool and smother whatever is feeding the fire. Water alone would spread liquid fires, that is, fires caused by inflammable liquids like oil, gasoline and kerosene because these substances can float in water. In fires like these, foam must be used.

What is foam made of?
It is a mixture of water with sodium bicarbonate and aluminum sulfate powder which forms carbon dioxide bubbles. The chemically inert carbon dioxide effectively blankets and smothers the flames.

How does a fire extinguisher work?
There are many different kinds. The one you see above contains pressurized carbon dioxide which is sprayed onto the flames.

A typical fire truck with: (1) water lines, (2) foam generator, (3) diesel engine, (4) radiator, (5) main drive shaft, (6) transmission and power take-off, (7) rear axle, (8) front axle, (9) steering column, (10) steering gear, (11) driver's seat, (12) foam nozzle controls, (13) directional foam nozzle, (14) "cherry picker," (15) filling port for foam tank, (16) filling port for water tank, (17) spotlights, (18) hose compartment, (19) control deck.

Is something added to water to help put out fires?
Certain products, called retardants, which modify the physical characteristics of water can be added to water. Some of these, called bathers, reduce the surface tension of the drops of water, improving their powers of penetration and diffusion. Others, called thickeners, form a kind of gel which makes the water viscose and adhere to combustible products. Chemical retardants absorb heat and deprive the fire of its fuel.

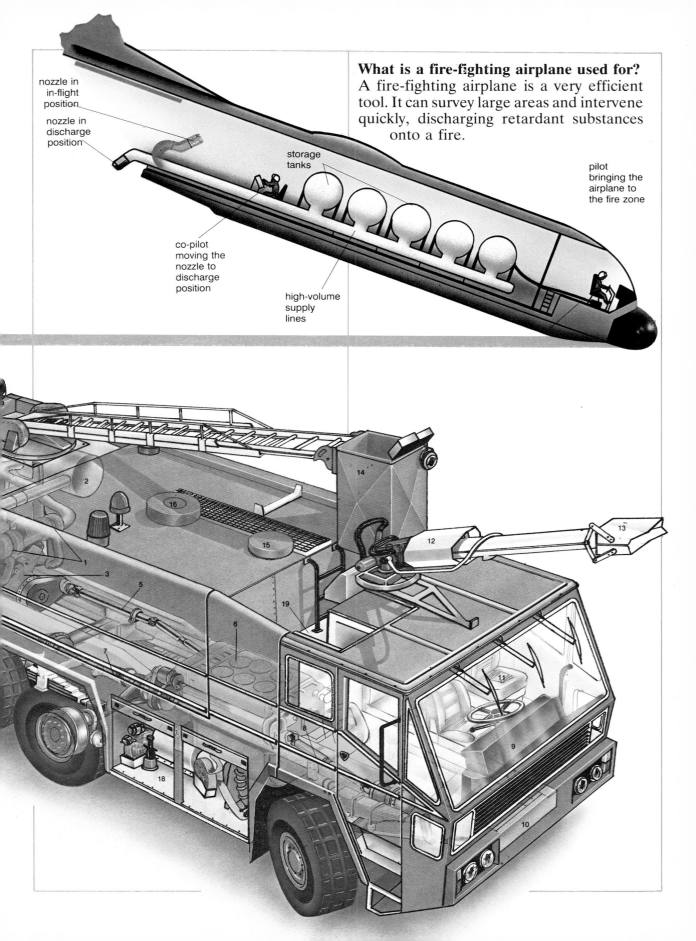

What is a fire-fighting airplane used for?
A fire-fighting airplane is a very efficient tool. It can survey large areas and intervene quickly, discharging retardant substances onto a fire.

nozzle in in-flight position

nozzle in discharge position

storage tanks

pilot bringing the airplane to the fire zone

co-pilot moving the nozzle to discharge position

high-volume supply lines

AMBULANCES

What is an ambulance?

It is a mobile medical unit equipped with everything needed for first aid and able to transfer the person treated – for a sudden illness, accident or medical problem – to a hospital or health-care facility where he or she can receive the most up-to-date technical assistance.

Are there ambulances that don't move?

In case of a natural disaster or war, fixed ambulances are set up. These first aid stations are next to or near disaster areas. When we use the word "ambulance," however, it is generally understood to mean a moving medical unit and, for this reason, it is sometimes called an ambulance car.

What characteristics must an ambulance have?

Above all, there must be enough space for the ambulance team to move easily, even standing up. There should be not only a wide rear door but also a side door, independent from the driver's cabin. The ambulance must have a two-tone siren and must be easily recognizable as an emergency vehicle. The minimal team is a driver and two stretcher-bearers, all of whom have completed a state-approved training course.

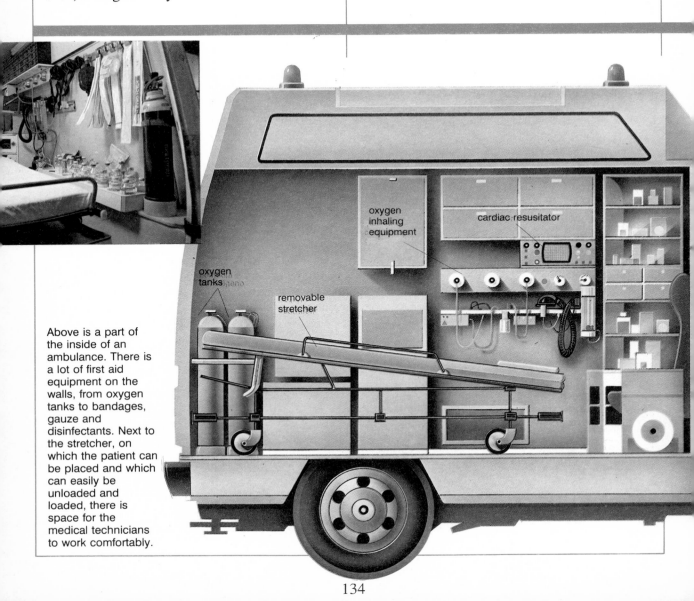

oxygen inhaling equipment

cardiac resusitator

oxygen tanks

removable stretcher

Above is a part of the inside of an ambulance. There is a lot of first aid equipment on the walls, from oxygen tanks to bandages, gauze and disinfectants. Next to the stretcher, on which the patient can be placed and which can easily be unloaded and loaded, there is space for the medical technicians to work comfortably.

What are the tools in the photograph below used for?

Ambulances are also equipped with tools for freeing an accident victim who is imprisoned or entrapped at the site of the accident.

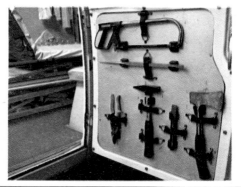

Is there a lot of equipment in an ambulance?

There's a portable first aid kit, material for immobilizing the limbs, the usual stretcher and another, called a "scoop stretcher," which is constructed of two vertical metal panels that fit together and can be used without moving the patient. There is a portable oxygen tank with a mask and a balloon with a mouthpiece, which is manually operated. Among the built-in equipment, there is also a water basin, an electric aspirator and an extra oxygen cylinder. It is possible to firmly fix the stretcher and a seat, called a cardiopathic chair, which is useful not only for those with heart problems but also for cases of rib fracture or serious respiratory difficulties.

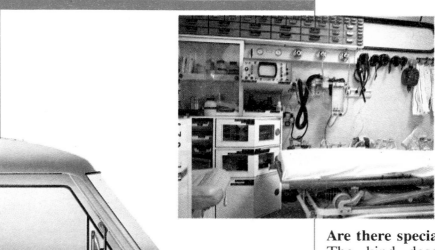

Special equipment within an ambulance: here you can see a heart resusitation monitor, used to give first aid to heart patients.

Are there special ambulances?

The kind described here is a typical ambulance. There are some, however, with special equipment for patients in particular conditions.

Can you give an example?

They range from units for transporting people who cannot move themselves to elaborate and prestigious mobile reanimation centers whose equipment corresponds to that found in an operating room. In the latter ambulance, it is possible not only to reanimate a seriously ill patient and to follow this patient's progress with a continuous electrocardiograph but also to do emergency operations. In this case, a anesthetist is part of the medical team.

AMPHIBIOUS VEHICLES

What does "amphibious" mean?
"Amphibious" has a Greek origin and literally means "double life." It is used in particular, for vehicles that can move both on land and in the water.

Are there military as well as civilian amphibians?
Most amphibians are constructed for military use. But there are also civilian types, used principally by sportspeople, explorers and firefighters, to give some examples. The vehicles for civilian use are called ATVs, All Terrain Vehicles.

What are military amphibians used for?
There are two main types of military amphibious vehicles, landing craft and command ships. Command ships are used to direct and supply land-sea operations. They carry helicopters, missiles, artillery, landing pads and communications equipment. Landing craft, like the example shown below, are used to carry men and supplies ashore. They can carry up to three tanks or eighty men who embark and disembark through a ramp at the bow.

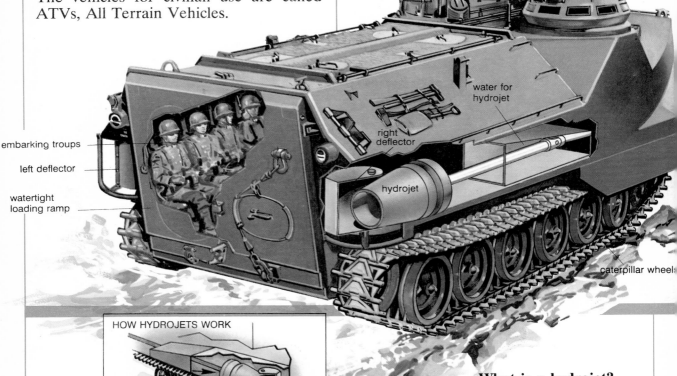

radio antenna

turret

water for hydrojet

right deflector

embarking troups

left deflector

hydrojet

watertight loading ramp

caterpillar wheel

HOW HYDROJETS WORK

water intake

water jet (forward movement)

water jet direction (deflected for a left turn)

left deflector

What is a hydrojet?
It is a special type of thruster for boats, used on amphibious vehicles. A motor activates a pump which receives water from a specially-designed opening and drives it through a tube with a nozzle at the end. The jet gives the push. A deflector (see the drawing on the left) directs the water jet in the desired direction. With this system, it is possible to maneuver even in deep water.

What are ATVs used for?
They are built to travel through marshes, swamps, lakes and rivers or over muddy, bumpy or steep terrain.

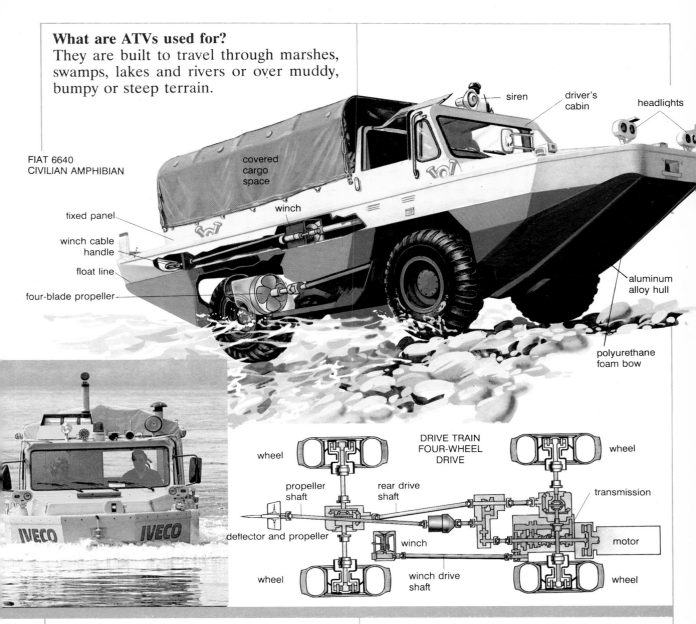

FIAT 6640
CIVILIAN AMPHIBIAN

siren

driver's cabin

headlights

covered cargo space

fixed panel

winch cable handle

float line

four-blade propeller

winch

aluminum alloy hull

polyurethane foam bow

IVECO   IVECO

DRIVE TRAIN
FOUR-WHEEL
DRIVE

wheel

wheel

propeller shaft

rear drive shaft

transmission

deflector and propeller

winch

motor

wheel

winch drive shaft

wheel

What do firefighters use amphibians for?
Firefighters use these vehicles for saving or giving first aid to flood victims.

How do these vehicles run on land?
ATVs have wheels or caterpillars. An ATV with caterpillar tread, like a tank or a bulldozer, runs on two treads.

Do they use hydrojet propulsion in water?
Not all do. They may also have propellers, like the one above, which is an amphibian equipped for use by firefighters. The propeller has four blades, and the direction (as

you can see in the drawing) is controlled by a rear deflector connected to the steering wheel.

What is a winch for?
The winch allows the amphibian to free itself if it gets stuck.

Isn't there a risk of leaks?
Yes and electrically-operated bilge-pumps are built into the vehicle for this reason. All of the mechanical groups shown in the drawings can be internally pressurized to avoid any water infiltration.

TRAINS

What is a railcar?
It is a carriage containing passenger seats, equipped with machinery and a driver's cabin. Railcars are often used on secondary lines where there are many stops and few passengers.

This is a carriage, or car, of a commuter train, a light structure having a body and a suspension with a large absorption capacity to guarantee a comfortable ride. The seats are on two levels, giving a larger seating capacity than a traditional train.

Are there many kinds of locomotives?
Yes, there are. An electric locomotive is easy to recognize because it has a kind of arm, the pantograph, on the top. During the journey, the pantograph touches the electric cable that runs above the train tracks and it takes the energy needed to feed the electric motors of the locomotive. Diesel locomotives have diesel engines, like buses and trucks. Turbine locomotives run on gas turbines. Diesel-electric locomotives have a diesel engine which, instead of transmitting movement directly to the wheels, feeds a generator which activates electric motors.

What is the train shown below?
It is a special train for laying tracks, with an engine, a work car and eight cars equipped with conveyor belts. A crane is moved along the cars to lift the ties and place them on the distribution chain (in red). The old ties are deposited on another chain (in blue). At point 1, the old ties are picked up, at point 2, the gravel bed is put in order and at point 3, the new ties are laid. The rails being replaced are somewhat bent along the length of the work beam. The same machine lifts the new rails from the side of the gravel bed, lays them in line and deposits the old ones by the side.

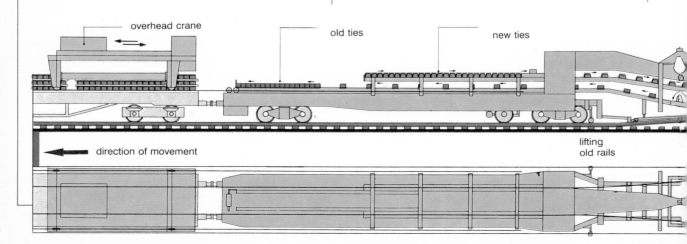

overhead crane

old ties

new ties

direction of movement

lifting
old rails

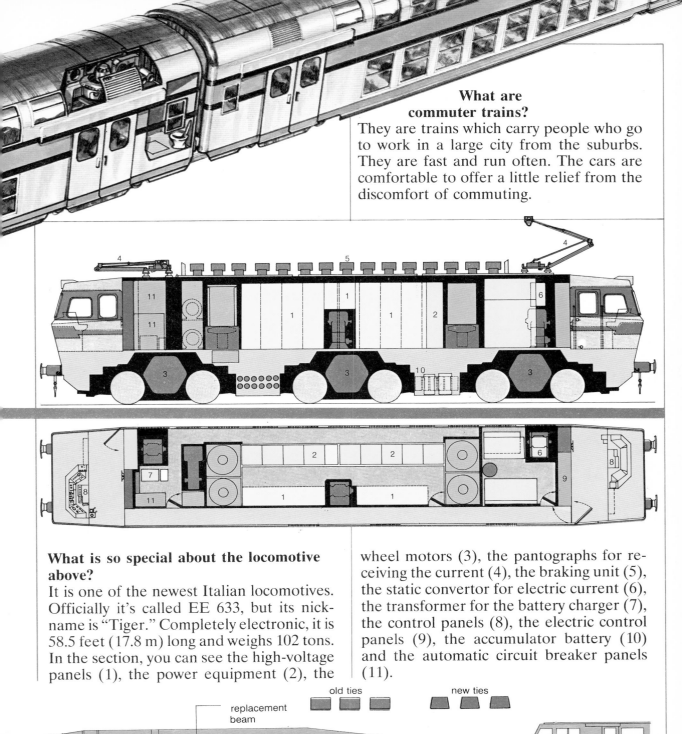

What are commuter trains?

They are trains which carry people who go to work in a large city from the suburbs. They are fast and run often. The cars are comfortable to offer a little relief from the discomfort of commuting.

What is so special about the locomotive above?

It is one of the newest Italian locomotives. Officially it's called EE 633, but its nickname is "Tiger." Completely electronic, it is 58.5 feet (17.8 m) long and weighs 102 tons. In the section, you can see the high-voltage panels (1), the power equipment (2), the wheel motors (3), the pantographs for receiving the current (4), the braking unit (5), the static convertor for electric current (6), the transformer for the battery charger (7), the control panels (8), the electric control panels (9), the accumulator battery (10) and the automatic circuit breaker panels (11).

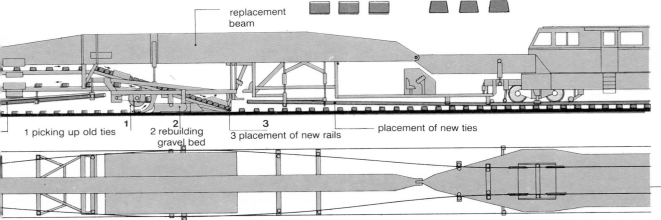

old ties   new ties

replacement beam

1 picking up old ties
2 rebuilding gravel bed
3 placement of new rails
placement of new ties

PUBLIC TRANSPORTATION

Did horse-drawn streetcars ever exist?

Yes, about the middle of the nineteenth century, this form of public transportation, in which a horse pulled passenger carriages along tracks, was very popular in European and American cities. It was, however, a very expensive form of transport due to the cost of buying and maintaining the horses. The horses were not able to cover long distances every day, and service was often endangered by illnesses. In 1872, for example, the Great Epizootic killed thousands of horses in the United States, halting service on many lines.

When were the first electric streetcars introduced?

As early as 1834, electricity - in the form of batteries - was used in urban transport. The use of batteries, however, proved to be too expensive and the first cablecars used steam power.

How did they work?

The cablecars ran on rails and were hooked onto a continuously moving cable below the surface of the street. A main powerhouse with steam engines pulled the cables along the various routes. The first cable line was built by a wire manufacturer, Andrew Hallidie, on hilly Clay Street in San Francisco. By 1890, many of the United States' cities had extensive cablecar systems.

A large streetcar, with a passenger capacity of 265, 205 standing and sixty seated. It is about ninety-six feet (29.20 m) long and has a maximum speed of thirty-seven mph (60 km/h). This vehicle consumes less energy than any other rail vehicle. It is driven by two engines, one at the front and one at the back. It is constructed using a technique that saves 6.5 tons of weight.

When were electric streetcars introduced?

The invention of the electric dynamo, or generator, in the 1870s made the development of the electric streetcar, or trolley, possible. From a dynamo, electric power could be transmitted along an entire rail route and could, then, run electric motors on the cars. In the beginning, the current ran in the rails and, later, through wires above the streetcar. A Belgian, Charles Van Depoele, perfected the trolley, an overhead wheel contact at the end of a pole, that ran below the power wire, held in position by a strong spring. Van Depoele installed the first electric streetcar, or trolley, in Montgomery, Alabama in 1886.

Are there any suspended trains?
Yes, the most famous one is in Tokyo (drawing at the top of the opposite page). They have no crossroads, traffic lights or other obstacles and can assure quick and regular passenger service.

Is this a monorail system?
Yes, it is. The most common monorail suspension system is the Alweg system in which the support wheels run along the top of the rail and the guide and balancing wheels run along the sides of the rail.

When was the first underground train built?
It was built in 1863 in London. Steam locomotives, burning coke and coal, were used.

What is the underground?
It is a network of electric trains that, partially or totally, run underground for use in urban transportation. It is advantageous, because it does not interfere with, or get slowed down by, urban street traffic. Since these systems are expensive to build, their presence is justified only in cities with high population density, where street traffic is frequently congested. (Above is the Milan underground.)

Are there also above-ground systems?
In most urban lines, the train begins underground and runs on elevated or ground-level tracks once it leaves the center of the city. It is much less expensive to build above ground, avoiding the cost of excavating underground tunnels.

Do undergrounds always run on electricity?
Modern undergrounds are electrically powered, like streetcars, except that underground cars pick up current from a lateral third rail instead of an overhead line.

LIFTS

How do lifts work?

All the different types of lifts are based, in one way or another, on the principle of counterweight, which means that the weight of one object can be used to balance the weight of another object. For example, if it's difficult to lift one hundred fifty pounds (70 kg) to the second floor by using a cable, it is much easier to attach a one hundred fifty pound weight to the other end of the cable and insert a pulley between the two weights. The second weight, the counterweight, counterbalances the first, which is then easier to lift. In a counterweight system, the weight on which the lifting force must be exerted is never equal to the total weight of the object but rather to the difference between the weight of the object and that of the counterweight.

When was the first lift installed?

Lifts of various types have existed for a long time. In the early eighteenth century in England, a steam engine was applied to the pulley to make the lifting operation easier. The first commercial lift for people was installed in a large New York City store in 1857 by Elisha Graves Otis.

How did it work?

It had a small steam engine which could pull it five floors in less than a minute. Reports from that era record the great success that this, and other similar lifts, had. Still they were very slow. If they were in use today, it would take more than fifteen minutes to reach the top floors of the highest skyscrapers.

Are lifts safe?

Lifts are surely the safest and most frequently used means of transport. In spite of the large number of users, there are fewer than a thousand accidents a year worldwide, largely involving people who get stuck in the closing doors.

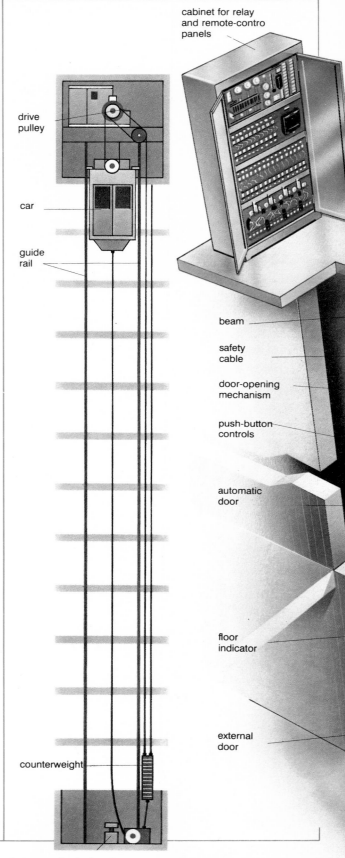

cabinet for relay and remote-contro panels

drive pulley

car

guide rail

beam

safety cable

door-opening mechanism

push-button controls

automatic door

floor indicator

external door

counterweight

shock absorber

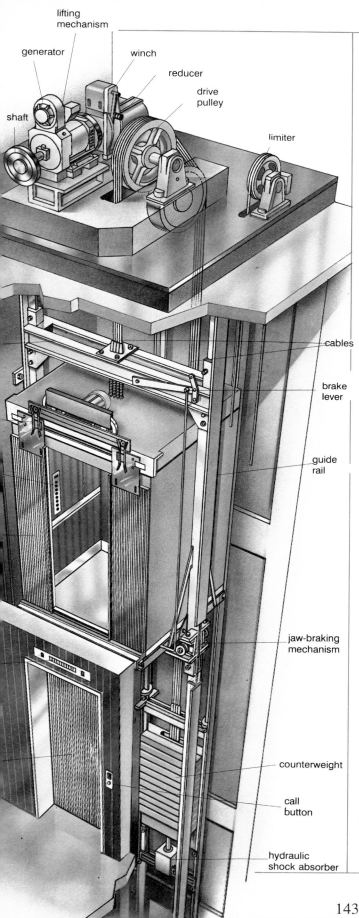

lifting mechanism

generator

winch

reducer

drive pulley

shaft

limiter

cables

brake lever

guide rail

jaw-braking mechanism

counterweight

call button

hydraulic shock absorber

Are modern lifts electric?

Yes, almost all modern elevators have electric motors to turn the pulley at the top of the shaft. A cable attached to the pulley makes the lifts go up and down at different speeds.

How is a modern lift made?

You can see one in the drawing between these pages. Note the double doors, one for the floor and one for the car, and an electric motor, or lifting mechanism, which includes an alternating current motor with a speed of up to four feet (1.20 m) per second and a direct current motor with a speed of up to 6.5 feet (2 m) per second. There is a reducer winch, a drive pulley and an electromagnetic brake. Above, on the left, you can see the cabinet which contains the movement and control panels with circulation in rectified current, able to function with voltage variations between + 10 and − 25 percent. All elevators use counterweights, and the diagram at the far left shows how the counterweight runs parallel to the car.

What is a turn regulator?

This is an improvement introduced by Otis to increase the safety of the lifts. It is a device which mechanically regulates the lift's speed, preventing it from going faster than a set speed.

How is it made?

A central shaft turns between two plates placed at either end of it, one of which is fixed and the other movable. Between the two plates, there are also two pairs of connection bars, with a joint in the center of each pair. There is a weight attached to each joint. When the shaft turns, it makes the weights turn as well, and they are driven to the outside by the rotation. The faster the rotation, the further they are driven. By moving, they drag the movable plate nearer to the fixed one until they come into contact. Since the movable plate cannot go farther than the fixed one, the weights cannot go farther from the shaft, and the shaft cannot turn quickly any more.

ESCALATORS

What are escalators?

Escalators are stairs in continual movement which allow you to go from one floor to another simply by standing on one stair. The widespread use of escalators began in the 1940s.

When were the first escalators built?

In the 1890s, two types of escalators were invented independently, one by Jesse Reno and the other by Charles Seeberger. The word "escalator" was introduced as a trademark of the Otis Elevator Company at the Paris Exposition in 1900, but it soon became the accepted name.

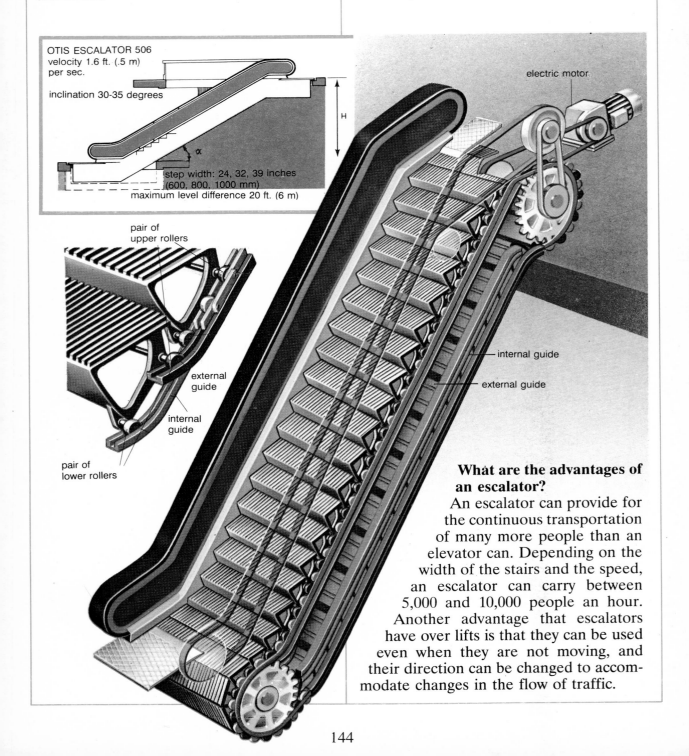

OTIS ESCALATOR 506
velocity 1.6 ft. (.5 m) per sec.

inclination 30-35 degrees

H

α

step width: 24, 32, 39 inches (600, 800, 1000 mm)
maximum level difference 20 ft. (6 m)

pair of upper rollers

external guide

internal guide

pair of lower rollers

electric motor

internal guide

external guide

What are the advantages of an escalator?

An escalator can provide for the continuous transportation of many more people than an elevator can. Depending on the width of the stairs and the speed, an escalator can carry between 5,000 and 10,000 people an hour. Another advantage that escalators have over lifts is that they can be used even when they are not moving, and their direction can be changed to accommodate changes in the flow of traffic.

What are moving sidewalks?

Moving sidewalks operate in the same way as escalators, but they do not have steps and, for this reason, can move on the level or on a slight incline (compare the drawing below with the escalator design on the opposite page). Moving sidewalks are used to facilitate movement in very extensive and crowded places, like airports or fairs.

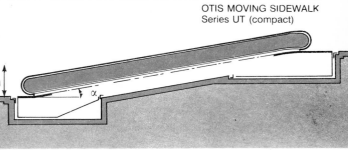

OTIS MOVING SIDEWALK
Series UT (compact)

velocity 1.6 ft. (0.5 m) per sec. – inclination 10-12 degrees

How does an escalator work?

Electrically powered by a motor on the floor above, an escalator is driven by two continuous, or closed, chains. The individual steps are attached to these chains by metal rods called step axles. The steps are supported by small rubber or nylon rollers on continuous steel tracks. The tracks control the steps, making them level out at the top and at the bottom, depending on the direction of the movement.

How are the chains attached?

At the bottom of the escalator, the chains lock into a toothed wheel and around wheels at the top connected to either end of the drive shaft, which is turned by the motor.

Is the chain mechanism similar to the one on a bicycle?

Yes, it is. When a cyclist pedals, the toothed wheel, to which the pedals are connected, moves the continuous chain which makes the toothed pinion on the rear wheel axle turn, turning the rear wheel. The drive shaft of an escalator acts like the pedals, turning the wheels, to which the chains are attached, around. As a result, the chains and the steps, connected to the toothed wheels, are set in motion.

How do the handrails work?

The handrails, which run in continuous loops in T-shaped guides situated along the tops of the escalator's side panels, are moved by other drive wheels at the same speed as the steps.

Where are escalators used?

You can find them in large stores, airports, train stations and, in general, places that have a heavy flow of pedestrian traffic which requires movement from one level to another. (At the top left of the page, you see an escalator in the Georges Pompidou Center, Paris.)

CABLEWAYS AND FUNICULARS

What is a cableway?
A cableway is any aerial means of transportation, for materials or for people, that uses a cable. Cableways mainly consist of towers which support a continuously-moving cable, from which the loads are suspended.

Is there more than one kind of cableway?
Yes, there is a specific type of cableway used for construction as well as another for the transporting of people.

What is a funicular?
A funicular is another means of cable transport which uses vehicles on rails to transport people to the tops of mountains.

What are the different types of cableways?
There are two principle types of cableways: monocable and bicable. The first type which uses only one continuous overhead cable, is supported by towers and vertical rollers and drawn forward like a giant pulley.

How does the bicable type work?
In a bicable system, there are two types of cable. One cable (or pair of cables), called the track cable, does not move. The track cable is similar to the tracks of a railroad – it supports the weight of the carts that hang from it on wheels. A second cable, to which the carts are fixed, called the haul cable, pulls the car along. One of the longest bicables ever built transported iron from

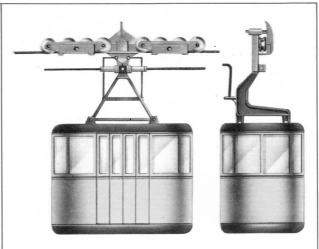

Kristinenberg to Boliden in Sweden – a distance of about sixty miles (96 km).

How large is a cableway car?
The car sizes vary greatly from one cableway to another. They go from cabins that can hold a few people to ones that can hold up to thirty, such as the above example, shown from the front and from the side. The illustrated car is made of aeronautic aluminum alloy, with a double eight-wheel trolley.

How are cableways that carry materials used?
These cableways, other than simply transporting loads, can pick up and drop off materials and, in this way, can be very helpful in construction. In some cases, they are used in mining. A cableway, built in 1926 in Monzone, Italy, lifted and transported marble blocks weighing more than twenty tons from quarries.

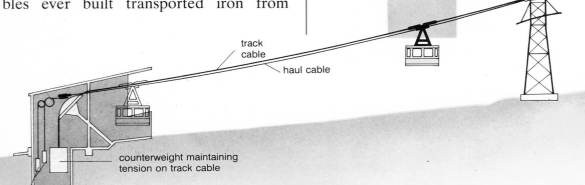

track cable

haul cable

counterweight maintaining tension on track cable

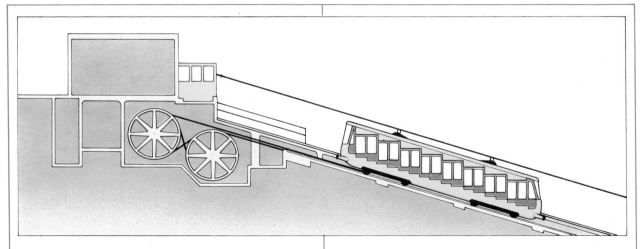

How does a funicular work?

As you can see from the drawing above, which shows the station at the top, the car runs on two rails, like a train. The movement is due to a large cable that is wrapped around a pulley, which you can see in the section of the station. If there is only one car, it is fixed on one of the cables while the other moves with nothing on it. On lines with many passengers, two cabins can be run on two adjacent rails or on only one which splits in the middle at the intersection.

What makes funiculars move?

Usually they are powered by electric energy. Since the cars are attached to a single cable, their weights balance and little power is required to move them. Some funiculars use no form of power other than weighting with water, filling their water tanks at the top and emptying them at the bottom. Since the top (descending) car is heavier than the bottom (ascending) car, it pulls the car from the bottom up as it moves down.

haul-cable winching drum

counterweight for haul cable

The drawing gives a complete idea of the "go and come" bicable type. This typical example of a tramway used for mass transit was built in 1976 to cross New York's East River from Manhattan to Roosevelt Island. Every day this bicable tramway transports between six and seven thousand people in its two cars. Another example of a bicable system is the cableway at Mont Blanc which boasts a section without ground towers 3,723 yards (2,831 m) long.

FROM CANOES TO OIL TANKERS

What boats were built in the past?

A dug-out tree trunk, invented by prehistoric people was the first means of water transport. From then on the shape and the means of propulsion have been constantly changing. Below you can see typical vessels from different historical periods. The Phoenician bireme was a warship from the seventh century, BC, with two banks of oars. The Roman trireme, a warship from the second century BC, with a warning tower and boarding gangway, could do five knots per hour. The Draker was a tenth-century Viking ship with a cloth covering and shields along the sides to protect the crew and the rowers. The Caravel, a three-mast Spanish ship from the sixteenth century, was about twenty-five yards (23 m) long. The Clipper, a slender and light three-mast ship made in the U.S.A. around 1820, could reach twenty knots per hour.

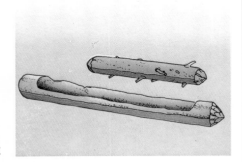

DUG-OUT TREE TRUNK

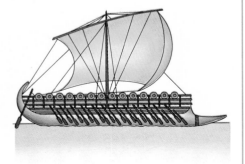

PHOENICIAN BIREME

ROMAN TRIREME

VIKING SHIP

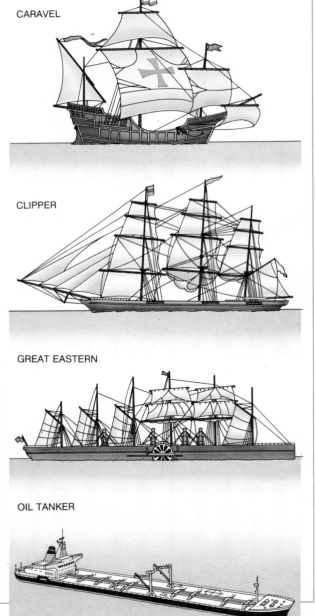

CARAVEL

CLIPPER

GREAT EASTERN

OIL TANKER

What was the Great Eastern?

The Great Eastern marked the passage from sail to steam power. It was two hundred twenty-nine yards long (210 m) and carried 4,000 passengers in addition to four hundred crew members.

What is an oil tanker?

It is a ship whose hold is divided into compartments for the transport of oil. Since the end of the 1960s, this type of ship has undergone enormous change. A newly-built supertanker may even weigh more than 500,000 tons.

What is the ship pictured above?

It is a ship specially constructed for container transport. These large containers, with a standard size and shape, can be easily loaded onto a variety of carriers (trucks, trains, ships, planes). In the picture, you see the steam propulsion equipment which drives this 30,000 ton giant with 40,000 hp.

STEAMBOATS

What kind of boat is shown on these pages?
This is a modern tour boat, an exact replica, both in shape and propulsion system, of the traditional riverboats. This boat, however, has modern conveniences and instruments.

How did riverboats work?
Riverboats with paddlewheels were the first boats on which experiments with steam engines were done at the beginning of the nineteenth century.

Who made the first riverboat?
The prototype of the riverboat, which was also, as it happens, the first steamboat in history, was made by a French aristocrat, Marquis De Jouffroy. It navigated a short tract on the Saone River near Lyon in 1783. But De Jouffroy's invention was forgotten after a while and it wasn't until twenty-four years later that the steamboat became an important regular means of transport.

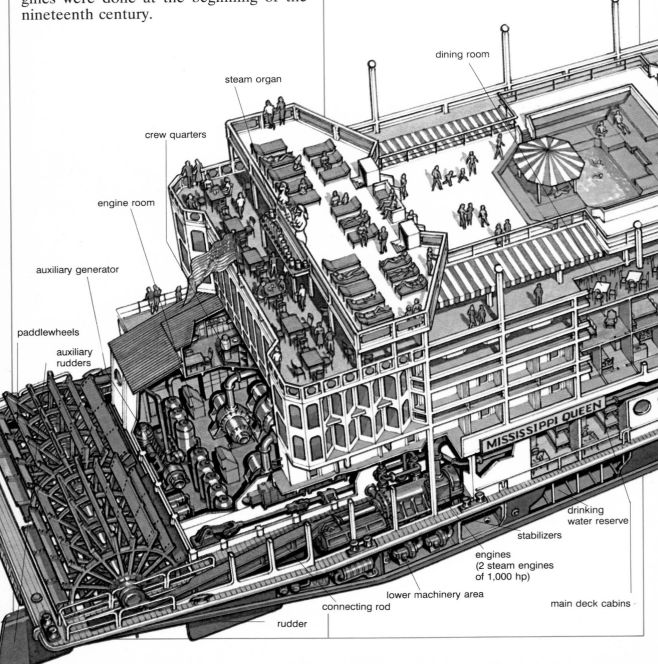

steam organ

dining room

crew quarters

engine room

auxiliary generator

paddlewheels

auxiliary rudders

MISSISSIPPI QUEEN

drinking water reserve

stabilizers

engines
(2 steam engines
of 1,000 hp)

lower machinery area

connecting rod

main deck cabins

rudder

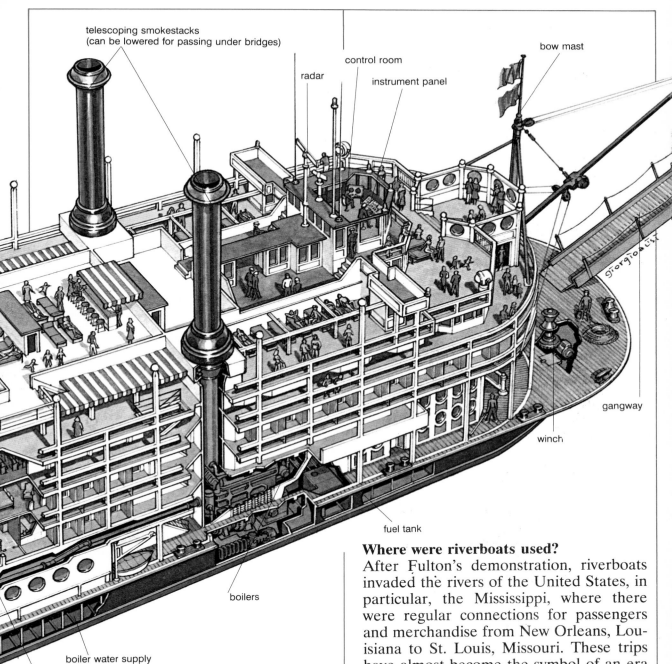

telescoping smokestacks
(can be lowered for passing under bridges)

control room

radar

instrument panel

bow mast

Giorgioaus

gangway

winch

fuel tank

boilers

boiler water supply

cabins

What is the Clermont?

The Clermont is the name of the 164 foot (50 m) boat built by the American inventor, Robert Fulton (1765-1815), which made steam navigation a real possibility. The Clermont, equipped with paddlewheels, made a great impression sailing down the Hudson River in April, 1807.

Where were riverboats used?

After Fulton's demonstration, riverboats invaded the rivers of the United States, in particular, the Mississippi, where there were regular connections for passengers and merchandise from New Orleans, Louisiana to St. Louis, Missouri. These trips have almost become the symbol of an era and are often used as the background for adventure films.

Are the steam engines of these boats the same as those of locomotives?

Yes, they are. The heat generated by the burning fuel transforms the water to steam, which is piped, under pressure, to a piston. In these boats, the movement of the piston is transmitted to the paddles.

OCEAN LINERS

What is a passenger ship?
It is a commercial ship equipped to transport people. Officially this title is reserved for ships which can carry more than twelve people. Ships of this type (see the picture for an example) usually are quite large, the larger ones exceeding 985 feet (300 m). They have all of the conveniences of a hotel in addition to particular safety and comfort features.

What is a deck on a ship?
A deck is a horizontal structure, extending from one side of the hull to the other, that carries cargo and passengers. The highest deck is called the upper deck. Structurally, the decks are made of a succession of beams which run horizontally. The placement depends on the width of the hull. Longitudinal beams, called carling, connect the horizontal beams. The covering is of metal or wood.

What is the captain's bridge?
The captain's or pilot bridge is where the principal information and control equipment for navigating the ship is located. Among the instruments that can be found on modern ships are the main compass, steering wheel, telegraph machine, radar screens, radio direction finder, electronic navigation instruments, marine chronometer and the ship's safety alarms.

smokestack

deck

deck

swimming pool

poop deck

engine room

stern

rudder

propeller

radar antenna

flag

captain's bridge

compass

prow

anchor

life rafts

hold

cabins

galley

Do the propellers makes the ship go forward?
Yes, by turning, they "spin into" the water, giving the ship the necessary thrust.

What are the propellers made of?
The propellers are usually made of bronze with manganese. They have very broad blades and can be very large indeed. The propellers of some supertankers, for example, have a diameter of 29.5 feet (9 m) and a weight of fifty-eight tons.

How many propellers does a ship have?
Cargo ships usually have one. Faster and more powerful ships, like passenger liners and many military vessels, have two. Even larger ships, like nuclear aircraft carriers, have four.

What propulsion system do modern ships use?
The most commonly-used method to move the ship through water is the marine propeller, activated by either an internal combustion engine or a steam engine connected to a chain of rods. In any case, these are engines derived from, and similar to, those used on land.

153

HYDROFOILS AND HOVERCRAFTS

What is a hydrofoil?

A hydrofoil is a motorized ship that is supported by a system of submerged wingline fins attached to the hull by strong, retractable struts. As its speed increases, it rises out of the water and slides along on the fins. In its movement, then, it meets only air pressure on the hull and water resistance on the fins.

What happens when the hydrofoil moves slowly?

The effect of the fins is negligible. In this case, it moves like any other boat. If the sea becomes very rough and the waves are higher than the "legs," the hydrofoil can always slow down, lower the hull into the water, and go ahead. Some hydrofoils can do this without retracting the legs and wings; other models can turn them around a scuttlebutt and extract the water, a method that is also useful for sailing in shoals. It can then proceed without using the wings.

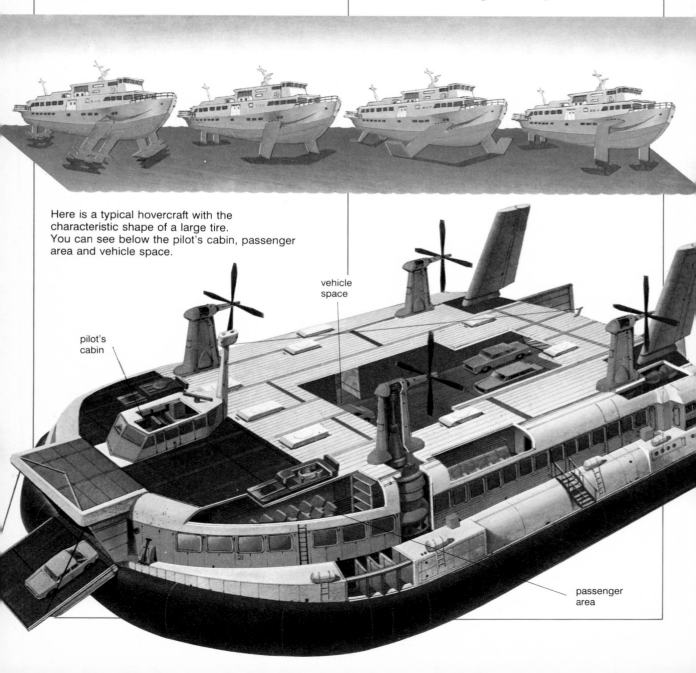

Here is a typical hovercraft with the characteristic shape of a large tire. You can see below the pilot's cabin, passenger area and vehicle space.

pilot's cabin

vehicle space

passenger area

What is a hovercraft?

It is a vehicle that can run on land or on water, sliding on an air cushion which can raise it from an inch or two to several feet above the water. The hull usually seems stocky, three times as long as it is wide, shaped like a large tire. To raise itself and to move, it usually has a pair of turboelectric engines.

In the photograph on the left, you can see a hydrofoil in motion, with the hull raised above the water. This is a Jetfoil by Boeing, with hydrojet propulsion. In densely populated areas, hydrofoils and hydrojets can be a valid alternative to the ever more congested land transit system. They can satisfy the strictest safety requirements and can be used for a variety of different purposes.

How does an air cushion work?

The operating principle is very simple: the air is taken in, compressed and pushed under the vehicle. A plastic or rubberized skirt keeps the air from escaping at the sides. In this way, an air cushion is formed, on which the vehicle rests and slides - even if it weighs several tons. Once formed, the air cushion is continuously fed by a flow of air toward the bottom and the outside. The amount of air to pump depends on the bumpiness of the land or the intensity of movement of the waves.

Are there many kinds of hydrofoils?

Four kinds are pictured at the left: there are those with stair wings, those a few inches from the surface of the water, those with V-wings, which are the most common, and those with underwater wings.

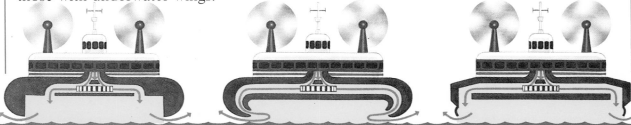

What is a hydrojet?

It is a ship with wings, similar to a normal hydrofoil, but equipped with jet propulsion. It can maintain a speed of forty-five knots even with waves thirteen feet high (4 m). The latest models have completely automatic piloting, based on sensors and gyroscopes.

What route does the air follow in a hovercraft?

That depends on the boat. The drawings above show the air flow in three types of hovercraft. The one on the right has a flexible border which keeps the vehicle at a certain height and helps it to overcome obstacles.

PORTS

How are ports built?

Until a few decade ago, ports were built in places where the coastline offered natural protection, which were not always the most economically practical locations. Today almost completely artificial ports can be built, even when the coastline is straight or backed by mountains.

How is an artificial port built?

A large area must be filled in and concrete wharfs and docks laid down. The bottom is dredged. Steel platforms are built off the coast for oil tankers. Obviously, ports like these are built not where it is possible but where it is convenient, for example, near places where raw materials are extracted or near the industries that convert these materials. Ports have undergone a notable development since World War II due to the increase in maritime traffic.

Are modern ports different from traditional ones?

At one time, cargo ships were more or less alike whereas today they are almost all specialized in a particular type of transport. Ports have, therefore, been adapted to this technological change. In almost all important sea and river ports, equipment has been installed for specialized incoming or outgoing cargo.

How is a port organized?

There is an area for miscellaneous merchandise, where the port has traditional equipment, and an area for bulk merchandise with the classical systems based on bucket cranes, metallic jaws which close over the material and unload it into a hold or onto a wharf. There is also new pumping equipment which takes in granular or powdery substances and unloads them directly into the places where they must go. There are loading and unloading areas for oil tankers, container ships, barges and so on.

Pictured is a dockyard scene showing the variety of freight-handling equipment available in a modern port: a crane on a floating pontoon, a traveling bridge crane with wheels, a bridge crane for loading and unloading containers and a series of silos. Other than multipurpose ports, ports equipped for a particular type of traffic, often adapted from pre-existing installations, are being built. These include ports for passengers, minerals, petroleum and fishing.

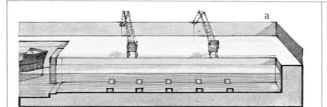

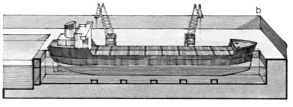

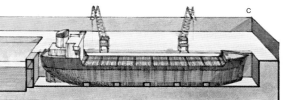

A dry-dock (a) is a facility used for repairing and maintaining ships in port: once the ship has gone in, the entry is closed (b), the water is pumped out and the hull emerges from the water (c). Nowadays floating dry-docks, large metal rafts formed from blocks, are often used (d). By first pumping in water (e), and then air (f, g) after the ship has entered, the dock floats, raising the ship into the air.

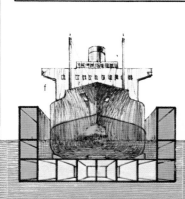

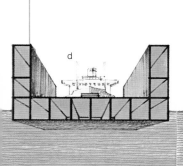

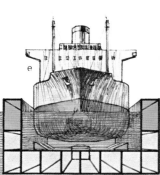

Are there special kinds of boats in ports?
Other than the police motorboats, towboats and fireboats, there are pilot-boats which carry specialized pilots whose job is to board ships and guide them, in place of their captains, into difficult port entrances.

SAILBOATS

Are there different categories of boats?

There are four main categories: canoes, which are paddled, rowboats, sailboats and motorboats. There are many subcategories of boats based on usage.

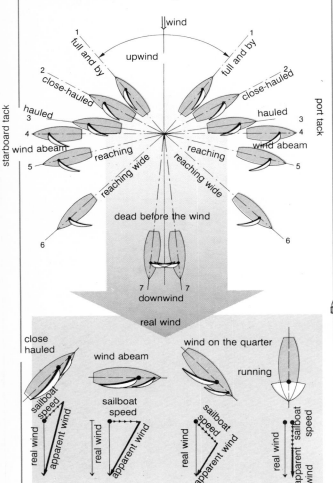

When were sails invented?

The first sails were made by the Egyptians who wove them from papyrus reeds and hoisted them on a single mast so that they could sail directly before the wind. The Phoenicians, who lived in the area of present-day Libya, improved on the Egyptian design, constructing small sailing vessels on which they traversed the entire Mediterranean, went as far north as England, and even rounded the Cape of Good Hope, centuries before the birth of Christ.

What is a yacht?

A yacht is a large sailboat, used mainly for amusement or sport, which can sail the high seas.

When was the first yacht built?

The first yacht was built in Holland in the sixteenth century. The Dutch had made a fast, light vessel with a single mast for chasing pirate ships. It was called Jacht-schiff, from Jagen, which means "to hunt" in Dutch. One of these yachts, the Mary, was given to Charles II, the English king, and the new concept of sailing for pleasure was born.

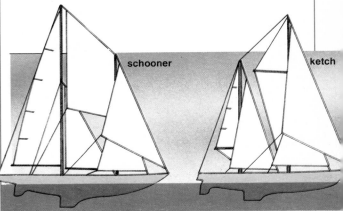

How does a sailboat move?

The drawings on the top left show a sailboat on different courses. The wind is coming from the top of the drawing, as shown by the arrow. The movements are called from starboard (with the wind coming from the left of the boat) and port (with the wind from the right). In the drawing in the middle on the left, the wind is blowing from the top of the drawing, and each drawing shows the same directional arrow, pointing downward. The small attached arrows, which always equal the same length and quantity, represent the speed of the boat, which is assumed to be the same in all the examples. The darkest arrow represents the apparent wind, that is, the on-deck measurement of the boat in motion taken by the wind gauge. You can see that this measurement is at its maximum for hauled sailing and at its minimum for running.

Are there different types of yachts?

Yes, there are various kinds of yachts in the nautical world. They are classified by a rating, a number calculated with a special formula that takes into account the various measurements of the craft and the sail surfaces. Some examples of yachts used for high seas racing are shown in the series of drawings in the middle of these two pages.

How is the hull of a sailboat made?

For centuries, the hull was made of wood. Nowadays it is more common to use plastic and synthetic materials, like fiberglass, Abs or Kevlar, or metals, like aluminum.

What equipment does a sailboat have?

The boats for one sailor usually have "cat" equipment: a single spanker (as in the drawing below), moved quite well forward. Other boats, besides the spanker sail, have a jib, which is white and triangular, like the spanker, but smaller, and a spinnaker, a light, large semispherical sail, which is usually very colorful, mounted on the prow.

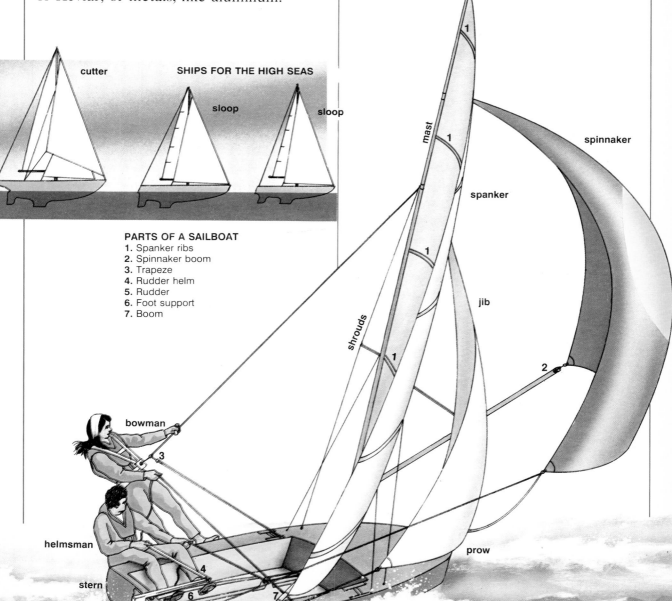

cutter

SHIPS FOR THE HIGH SEAS

sloop

sloop

PARTS OF A SAILBOAT
1. Spanker ribs
2. Spinnaker boom
3. Trapeze
4. Rudder helm
5. Rudder
6. Foot support
7. Boom

wind direction

mast

spinnaker

spanker

jib

shrouds

bowman

helmsman

stern

prow

MOTORBOATS

What is a motorboat?
It is a boat with a propulsion engine, used principally for pleasure.

When was the first motorboat made?
The first patent for a motorboat was taken out on October 9, 1886. This boat, designed by a German, G. Daimler, had a 1 hp engine and ran on petroleum.

For how long have motorboat races existed?
They have existed since the beginning of the twentieth century. Since then the speed of a motorboat has risen from 16.9 knots (19.53 mph or 31.42 km/h), set in the first Harmsworth Trophy Race in Ireland in 1903, to 247.7 knots (285.213 mph or 459 km/h) reached by a jet-propelled boat in 1967.

What materials are used for the hull?
The materials used for building a motorboat are the same as those used for a sailboat. Wood is still used, but with special resins and phenolic glues to increase its resistance. Marine plywood, made of several thin, even layers of the same high-quality wood, is used most often. Steel, which is used in large boats, must always be painted to protect it from rust and sea vegetation. Often the steel is also coated with resins. Light and resistant aluminum alloys are also used, but since they are not very pliable, they are not good for curved surfaces. Fiberglass, a new material, is more and more often employed; it has a technical and economic advantage in that it can be used in assembly-line production.

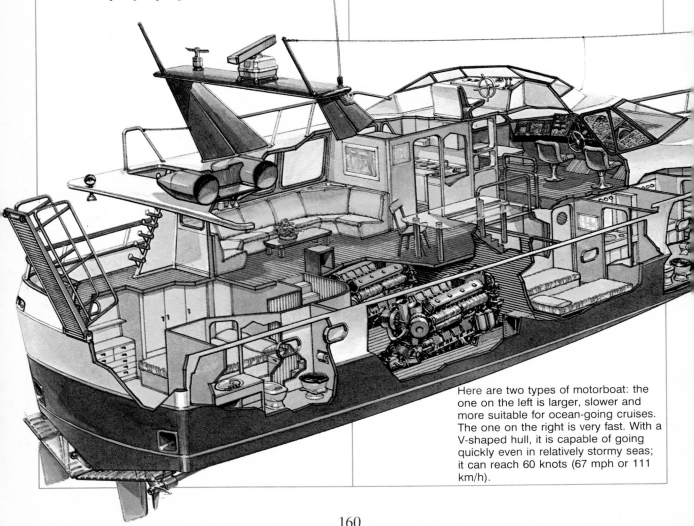

Here are two types of motorboat: the one on the left is larger, slower and more suitable for ocean-going cruises. The one on the right is very fast. With a V-shaped hull, it is capable of going quickly even in relatively stormy seas; it can reach 60 knots (67 mph or 111 km/h).

Are there many kinds of motorboats?
Yes, there are. They are usually classified as outboards, inboards, inboard/outboards, semi-cabin or cabin cruisers.

What does "outboard" mean?
The expression refers to the position of the motor; in this case, the motor is outside of the boat. An inboard, instead, is in the boat. An inboard/outboard has its motor inside and its foot outside. Outboard motors, using gasoline or a gasoline-oil mixture, are not very powerful. Inboard motor boats are rather small, having a length of ten to twenty feet (3-6 m).

Does the inboard/outboard offer any advantages?
It allows you to keep the motor repaired. The propeller can be lifted and turned, given the maneuverability of the foot.

Which motors are used as inboards?
They are almost always diesel motors, placed in the boat and joined to the propeller by an axle line. Inboard motorboats may be as long as thirty feet (9 m). The motors may have diverse horsepower and weight; they may reach 3,000 hp and weigh almost 12,273 pounds (5000 kg).

What are semi-cabin and cabin cruisers?
They are motorboats designed to give some comfort to those who go boating for pleasure. The semi-cabins usually have the driver's seat in the open with a couple of beds and a small bathroom in the prow. Cabin cruisers, depending on their length, which may reach about forty feet (12 m), can house three to seven people. They usually have a bathroom and kitchen as well.

What is a motor sailer?
It is a boat that can navigate either with a sail or a motor. Motor sailers are designed to take advantage of strong winds for sailing and to use the motor when the winds are low.

What is a motor yacht?
It is a boat that is halfway between a cabin cruiser and a yacht; in reality, it is a ship.

How long is a motor yacht?
A motor yacht may be between thirty-three and ninety-nine feet (10-30 m) long. These vessels can offer comfortable accommodations for several people for quite a few days. They have inboard motors.

BATHYSCAPHES

What is a bathysphere?

Based on the hot air balloon, a bathysphere is one of the first methods used for deep-sea diving. It was invented for scientific purposes: to observe the sea bed, understand the formation of the earth's crust and learn about marine life.

How is a bathysphere made?

The bathysphere consists of a pressurized spherical cabin which houses the crew and some of the equipment (comparable to the basket of a balloon) and a float which supports the cabin (comparable to the balloon itself).

Are bathyspheres very large?

Yes, they have a considerable size and weight. In fact, they cannot be loaded on an accompanying ship but must be towed to the diving site. This limitation complicates putting them into operation, especially when the sea is rough.

What is a bathyscaphe?

A bathyscaphe can be seen as a small submarine strengthened to resist strong pressure and able to move independently under water since it is not connected to another vessel. Bathyscaphes have taken the place of bathyspheres in underwater exploration.

When was the first bathyscaphe built?

The first successful immersion of the bathyscaphe, invented by Auguste Piccard, was with the model FN RS-2, which dove to a depth of 4,920 feet (1,500 m) off the coast of Cape Verde in 1948. This first dive was done without a crew. This bathyscaphe was then rebuilt by the French Navy and used, until 1958, for about sixty dives. It was pushed to a depth of 13,448 feet (4,100 m) off the coast of Dakar without experiencing any difficulties.

How is a bathyscaphe made?

It consists of a hull filled with gasoline and a pressurized spherical cabin for the crew. The hull allows for vertical movement by regulating the floating force and for horizontal movement.

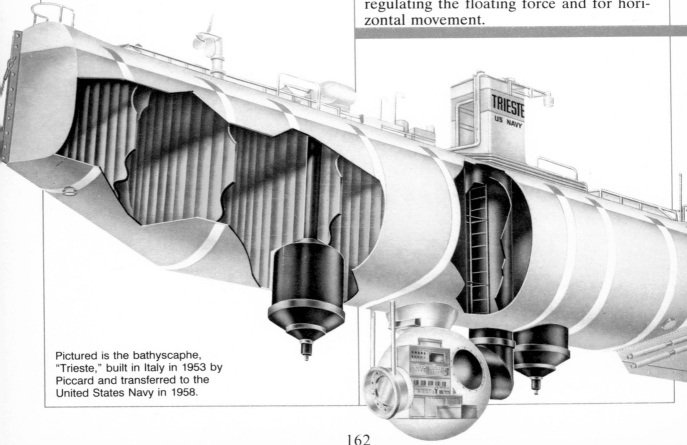

Pictured is the bathyscaphe, "Trieste," built in Italy in 1953 by Piccard and transferred to the United States Navy in 1958.

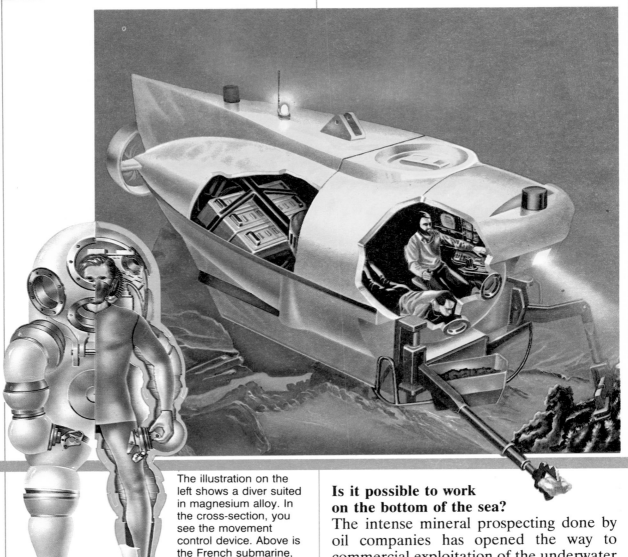

The illustration on the left shows a diver suited in magnesium alloy. In the cross-section, you see the movement control device. Above is the French submarine, the Nautilus, used for observations of the ocean bed.

Does a bathyscaphe float on the surface?
Yes, it does. It is not designed to withstand high underwater pressures. To counterbalance the external pressure, sea water is allowed to penetrate the bottom of the hull. The gasoline, which is lighter than water and does not mix with it, floats in the upper part. Iron blocks, attached to the outside by electromagnets, provide ballast for the hull. When there is not enough power, the ballast automatically detaches and the bathyscaphe rises to the surface. In case of breakdown, this is the bathyscaphe's main safety device.

Is it possible to work on the bottom of the sea?
The intense mineral prospecting done by oil companies has opened the way to commercial exploitation of the underwater environment. Often, to build and maintain underwater facilities, it is necessary to work at considerable depths. This work can be done using special cabins, modern versions of the diving bell.

Are there suits for diving to great depths?
The drawing on the left shows a special suit made of magnesium alloy which allows a person to work in normal conditions up to a depth of 1,642 feet (500 m). Equipped with a long-lasting oxygen tank, it is mainly used for work in the petroleum field. It is called Jim, after the person who tested it.

SUBMARINES

When did the first underwater boat appear?

It was in 1776, during the American Revolution. It was David Bushnell's American Turtle, activated by only one person turning a handle which controlled the rotation of a propeller. In 1899, a Frenchman, Maxime Laubeuf, had the idea of constructing a double-hulled vessel and of having it descend, ascend or remain stable by filling or emptying the space between the two hulls (the ballast box) with water. He called it the Narval.

How do submergible vessels move?

A classic submergible vessel, underwater and on the surface, is powered by a storage battery which, when the vessel is on the surface, is recharged by a generator with a diesel motor. Since the battery has a relatively limited capacity, the submergible must often rise to the surface, where it is easily visible.

What is a snorkel?

A snorkel is a kind of long tube installed by the German Navy on its submergibles during the second world war to reduce vulnerability. The snorkel allowed the diesel motor to "breathe" in water up to about fifty feet (15 m) deep. It was later adopted by all navies and installed on all underwater vessels, including nuclear submarines. The snorkel is somewhat inconvenient as it requires the submergible to stay near the surface where it may be discovered, either by sight or by radar.

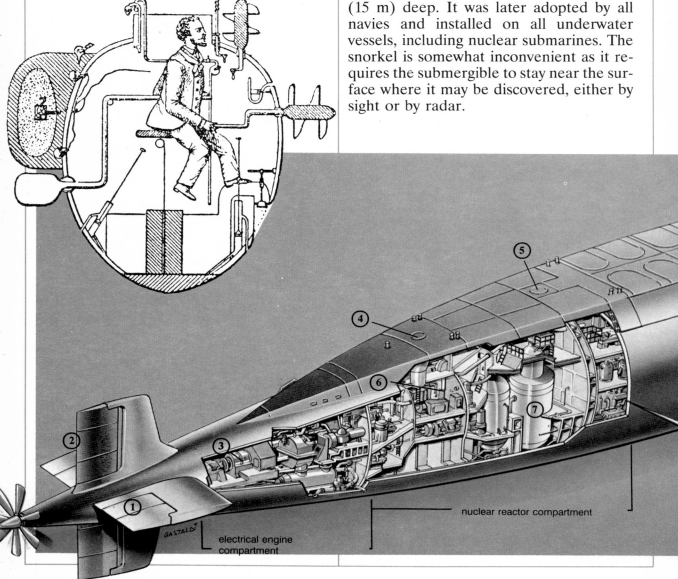

GASTALDI

electrical engine compartment

nuclear reactor compartment

What is the difference between a submergible and a submarine?

For a long time, underwater vessels were conceived of as surface ships capable of submerging, that is, as submergibles. The conception was, in fact, justified since their underwater performance (speed, maximum depth, length, means of navigation and detection) were very limited. With the coming of nuclear power, submarines have truly earned that name, since they are now more suited for underwater than for surface operations.

What are the characteristics of nuclear submarines?

Their shape allows them to move with a relatively low consumption of energy and a very low noise level. Their range of action is practically unlimited, and their ability to travel long distances is no longer a technical problem but depends, instead, on the endurance of the crew.

How does nuclear propulsion work?

Nuclear reactors produce heat. The water vapor that results makes the turbines rotate, giving mechanical energy to the propeller shafts and to the electric generators. These generators supply electricity to all the auxiliary equipment used for the operation of the submarine and for the crew.

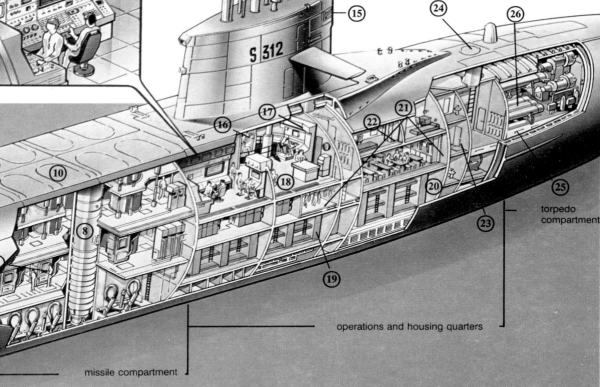

missile compartment

operations and housing quarters

torpedo compartment

This is a cross-section of a nuclear submarine. You see the horizontal rudders (1), the vertical rudder (2), the spare diesel-electric motor (3), the rear exit hatch (4), the rear entry hatch (5), the control room (6), the nuclear reactor (7), the M-4 missile in the launch bay (8), the missile launch control center (9), the hatch for the missile launcher (10), the periscope (11), the radar (12), the radio antenna (13), the snorkel or air intake (14), the surface control room (15), the periscope viewfinder (16), steering apparatus with viewing screen (17), the charthouse (18), the NCO's quarters (19), the crew quarters (20), the dining room and kitchens (21), the officers' quarters (22), the bathrooms (23), the fore exit hatch (24), the torpedo launch site (25), the additional torpedoes (26).

AIRCRAFT CARRIERS

When did the idea of aircraft carriers make its appearance?
Right from the beginning of aviation, there were those who dreamed of having airplanes take off from ship runways. These runways were first at the stern and later at the bow before evolving into a single runway from stern to bow.

When did the idea begin to spread?
Ships that were modified to allow for take-offs and landings could no longer also carry high-powered artillery. For this reason, not until airplanes could offer something more interesting than the artillery, especially in enlarging the range of action and the military power, did the navies of different countries begin to build ships designed to carry and use airplanes.

Where were the first aircraft carriers built?
The very first aircraft carriers were built in Great Britain beginning in 1916. France built its first one in 1920. The United States began to equip itself with carriers in 1925 while Japan started construction of these new ships in 1921.

Where are aircraft carriers most used?
In the years following the second world war, the United States Navy developed the largest number of these vessels. The first of these great aircraft carriers, the Forrestal, entered into service in 1955, followed by the nuclear carrier, the Enterprise, at the end of 1961. Navies of other countries have chosen different solutions, selecting for their needs ships equipped for vertical take-offs or helicopter carriers, a new concept that moves away from the traditional type of aircraft carrier.

What are the main parts of an aircraft carrier?

You can see them clearly in the drawing below. From its command station (10), an officer orders the launching device (3) to prepare the catapult (1) on the runway (6). Protected by the safety net (4), crew members can later recuperate the catapult sling in (2). At the beginning of the runway, which is flanked by transmission antennas (5), a deflector (7) pushes the jet steam forward. The landing, on the stern deck, is aided by arresting cables (11). The planes are brought out of the hangars (8), by elevators (12), and placed on the deck (13). The ship, powered by nuclear reactors (9), has numerous radar systems for surface and air surveying (14), an artillery radar (15), and approach-control radar (19). Below the admiral's deck (17), is the navigation deck (16), while the runways are supervised from the aviation commander's deck (18). The crew quarters, the workshops, the fresh water tanks, the fuel and the ammunition are on the eight floors under the flight deck, the lowest of which are below the water line.

How many planes can an aircraft carrier hold?

That depends on the size of the ship and on its dislocation. The number of aircraft on board approximates the number of thousands of tons of the ship's weight. The aircraft carriers in service have a displacement of between 20,000 and 100,000 tons. Therefore, the number of aircraft on board varies from about twenty to one hundred. Different aircraft, for example, interceptor planes, reconnaissance aircraft, fighter planes and connection and service planes and helicopters, all have different dimensions.

EARLY AIRPLANES

What were the first flying machines?

They were different types of balloons (see the drawings on the opposite page), shells which contained a lighter-than-air gas and which, for this reason, were able to rise off the ground.

Are airplanes heavier than air?

Yes, they are. People had long imagined that heavier-than-air machines could fly. Leonardo da Vinci deduced that since birds, whose weight is greater than the weight of air, use their wings to fly, the basic principle to explore for artificial flight was in the wings. The first airplane studies were made by Sir George Cayley at the beginning of the nineteenth century.

When did the first airplane fly?

After many attempts, two American brothers, Wilbur and Orville Wright, succeeded in flying. On December 17, 1903, they brought a plane that they had made to Kitty Hawk, North Carolina. Orville put himself at the controls, and the little engine, which moved two large propellers at the back of the plane, was started. The plane, helped by the counterweight, rose into the air. The event is shown in the drawing below.

How long did the first flight last?

The plane stayed in the air for twelve seconds, covering about one hundred eighteen feet (36 m), at a very low altitude, about three feet (1 m) off the ground. Nevertheless, the goal had been reached!

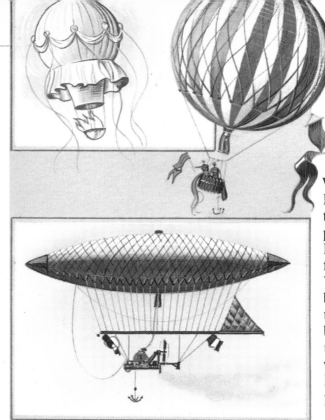

What was the Wright Brothers' first plane called?

It was called Flyer. It had a wingspread of 40.2 feet (12.25 m) and a length of twenty feet (6.12 m). It was eight feet (2.43 m) high and weighed 737 pounds (335 kg).

What followed the Wright Brothers?

Immediately after, the Englishman, Horatio Phillips, built a steam-propelled multiplane. Then the Brazilian, Alberto Santos-Dumont, made the first real airplane which flew on September 13, 1906 in Europe. The Voisin Brothers, from France, succeeded in building a floating biplane, the precursor to the seaplane. In America, the Wright brothers had no competitors for some time until Glenn Hammond Curtis, a young and willing builder-pilot, began to enter the limelight, winning two important aviation prizes in 1909.

PROPELLER PLANES

Why does an airplane fly?

It flies because it has wings, in particular, it has wings of a certain shape, as you can see in the drawings on these two pages. The shape of the wings is very important because it is the element that produces the lift, or the push upward, which supports the plane at a given speed.

Sopwith Camel

The heavy airplanes of the early 1920s experimented with improved wing shapes which were wider and better constructed.

TURBOPROP ENGINE

How does lift work?

Many people think that planes stay in the air because they are supported by the push from the air below the wings, but that is not true. A wing has a rounded front edge, the leading edge, and a honed back edge, the trailing edge. Because of this shape, the air that passes over the wing has to travel farther, and, therefore, faster, than the air that passes under the wing. The faster moving air becomes more rarified and exercises less pressure than the air under the wing, which creates a kind of sucking effect, pulling the wing upward. This sustained thrust which gets the plane into the air is the lift.

Vickers Valentia

The entire leading edge had a hard covering but the thickness was still not enough to eliminate wind bracing.

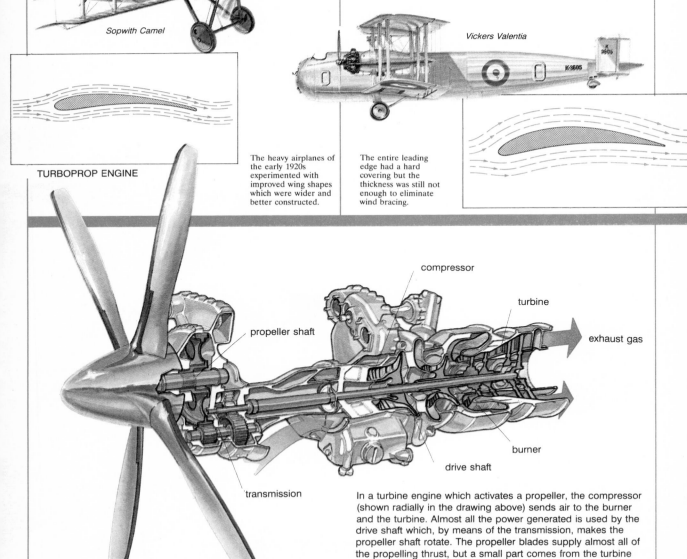

compressor

turbine

exhaust gas

propeller shaft

burner

drive shaft

transmission

In a turbine engine which activates a propeller, the compressor (shown radially in the drawing above) sends air to the burner and the turbine. Almost all the power generated is used by the drive shaft which, by means of the transmission, makes the propeller shaft rotate. The propeller blades supply almost all of the propelling thrust, but a small part comes from the turbine exhaust.

propeller blade

Is the design of the wings very important?

The wings are the part of the airplane that interact with the air to make the plane fly. Even beginning with the Wright Brothers, it was understood that, to obey the lift principle, it was necessary to study the curvature of the dorsal (upper) and ventral (lower) surfaces. In a short time, some basic wing types began to emerge (see the drawings on these pages). Broad wings with curved leading edges were the best for climbing, that is, they generated strong lift, but they did not allow for high speeds. Flight control was improved by moving the widest point to the back along the rigging. Thinner wings, which allowed for higher speeds, made the plane difficult to maneuver and required considerable ability.

What kinds of engines can a plane have?

A plane may have a piston engine, which makes a propeller rotate, or a reaction engine, a jet, with compressor and turbine.

How does a propeller plane work?

Airplanes fly because of the action-reaction principle. In the case of a propeller, the action exercised by the propeller drives back an air mass giving it a determined acceleration. As a reaction to this force, there is a force applied in the opposite direction, a push on the mass of the airplane.

Is the shape of the fuselage important?

Yes, it is. This was not really understood until the development of the wings was already quite advanced, at the beginning of the 1920s.

The Trimotor, below, from the end of the 1920s and early 1930s, introduced a thick shape that had strong lift but was not fast. For the first time, it became possible to use the wing covering to strengthen the wings. Finally wind bracing disappeared. But the aerodynamics of the fuselage was still underdeveloped, especially the wing-fuselage connection.

Ford Trimotor

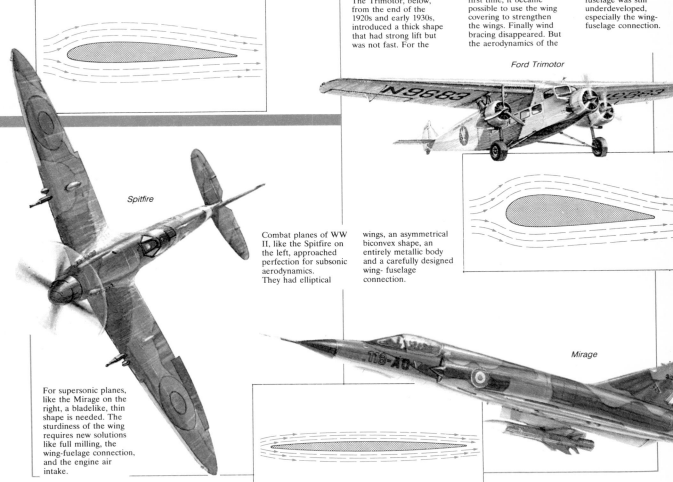

Spitfire

Combat planes of WW II, like the Spitfire on the left, approached perfection for subsonic aerodynamics. They had elliptical wings, an asymmetrical biconvex shape, an entirely metallic body and a carefully designed wing-fuselage connection.

For supersonic planes, like the Mirage on the right, a bladelike, thin shape is needed. The sturdiness of the wing requires new solutions like full milling, the wing-fuselage connection, and the engine air intake.

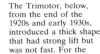

Mirage

JETS

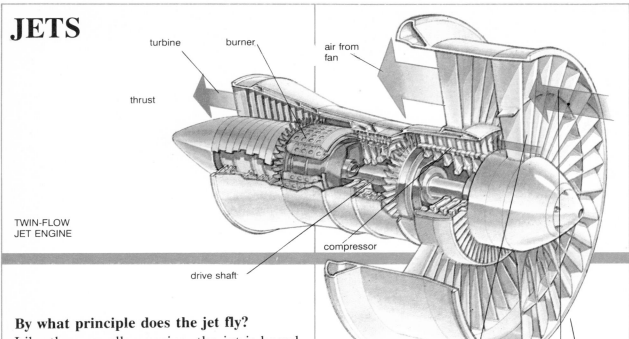

turbine   burner   air from fan

thrust

TWIN-FLOW
JET ENGINE

drive shaft

compressor

air intake
to turbofan

air intake
to compressor

fan

By what principle does the jet fly?

Like the propeller engine, the jet is based on the action-reaction principle. The action is exercised by a gas turbine which expels a mass of air and combustion gas through a exhaust nozzle, producing a strong acceleration. As a reaction, the mass of the plane is pushed in the opposite direction.

What is a turbofan?

A twin-flow jet engine is shown in the drawing at the top of the page. Part of the intake air which feeds this type of engine goes into the compressor, the burners and finally the turbine, producing a large part of the thrust. But the rotation of the turbine activates the main drive shaft which, other than keeping the turbine rotating, also makes the large fan turn. This fan sucks the air and drives it backward after having somewhat compressed it. This is the air that gives the main thrust to the engine.

Does this type of engine have important advantages?

With this design, the engine output is considerably greater than that of a pure jet engine. Moreover, the exhaust air from the large fan, having a low pressure compared to that of the turbine jet, decreases the jet noise.

What are the control surfaces?

They are the surfaces that regulate the movement of an aircraft along its longitudinal, transversal and vertical axes.
They are shown in the drawing on the right for a typical modern airplane.

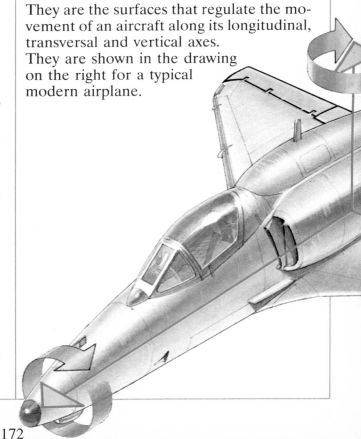

172

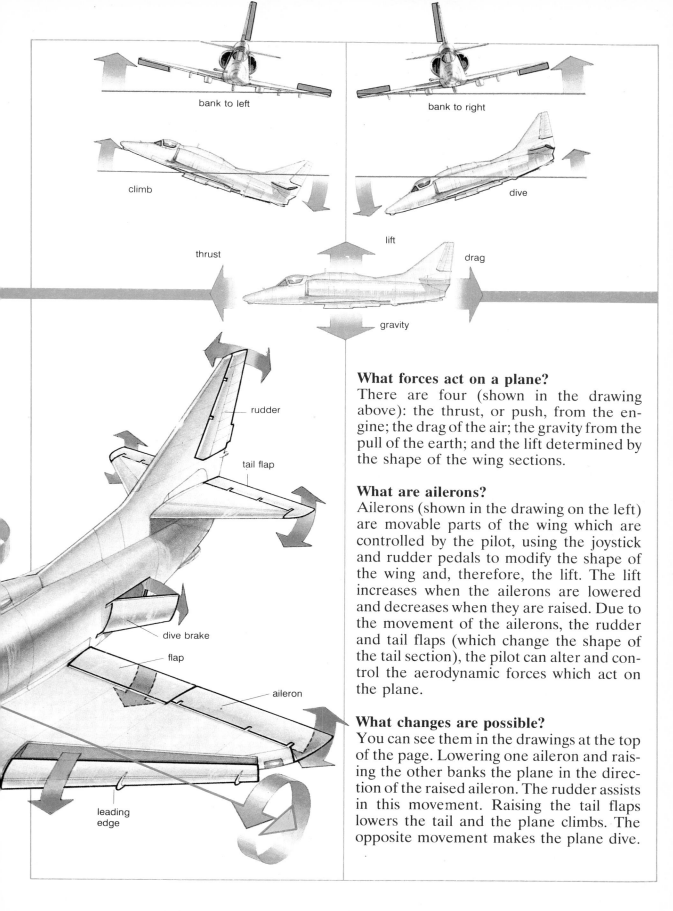

bank to left

bank to right

climb

dive

thrust

lift

drag

gravity

rudder

tail flap

dive brake

flap

aileron

leading
edge

What forces act on a plane?

There are four (shown in the drawing above): the thrust, or push, from the engine; the drag of the air; the gravity from the pull of the earth; and the lift determined by the shape of the wing sections.

What are ailerons?

Ailerons (shown in the drawing on the left) are movable parts of the wing which are controlled by the pilot, using the joystick and rudder pedals to modify the shape of the wing and, therefore, the lift. The lift increases when the ailerons are lowered and decreases when they are raised. Due to the movement of the ailerons, the rudder and tail flaps (which change the shape of the tail section), the pilot can alter and control the aerodynamic forces which act on the plane.

What changes are possible?

You can see them in the drawings at the top of the page. Lowering one aileron and raising the other banks the plane in the direction of the raised aileron. The rudder assists in this movement. Raising the tail flaps lowers the tail and the plane climbs. The opposite movement makes the plane dive.

HELICOPTERS

Who invented the helicopter?

In 1483, Leonardo da Vinci designed a spiral-shaped wing which should have been able to spin in the air, lifting the air machine to which it was attached. Lacking an adequate engine, this rotating wing never flew, but it is the ancestor of the modern helicopter which flies by using a rotor, with two or more blades, activated by a motor.

What is the main difficulty with helicopter flight?

The rotor blades do not move at a constant speed. Because they are spinning in a horizontal plane, the blades on one side, moving forward into the wind, develop more lift than those on the other side, which are moving downwind. This creates an imbalance in the lifting force, which has to be adjusted to control the helicopter.

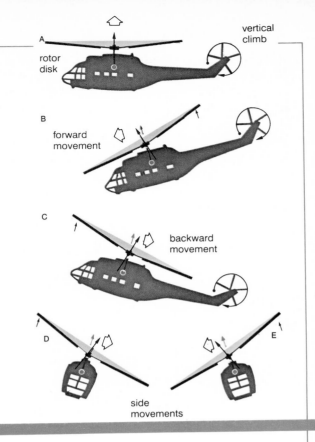

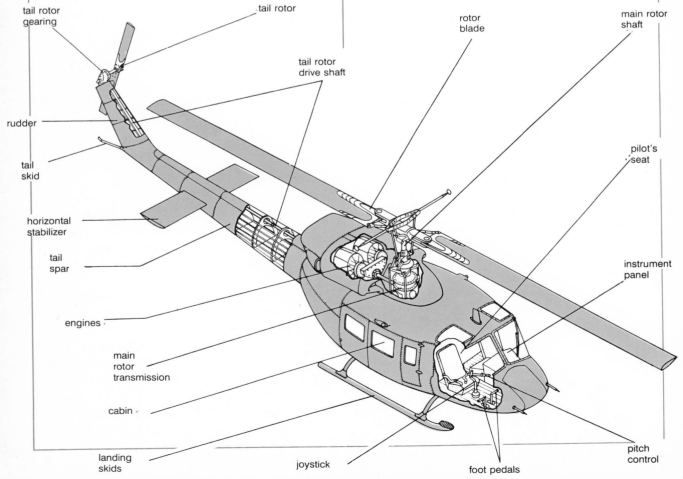

How was the rotor problem solved?

It was solved by designing rotor blades that changed their speed of rotation at every turn and, therefore, exercised variable lift depending on their position. For example, if the blades have greater lift when they are at the back, the helicopter receives a push from behind and goes forward.

How is a helicopter controlled?

A helicopter has three basic controls (see the drawing at the bottom of the opposite page): the pitch control which regulates vertical climb (A, in the drawing on the left), and the stationery control, which can simultaneously vary the pace of all the blades; the joystick, which controls the forward, backward and side movements, tilting the rotor disk plane (B, C, D, E); and the foot pedals, which regulate the tail rotor thrust, turning the helicopter to the right or left.

How does the tail rotor work?

By spinning, the tail rotor produces a thrust which tends to make the helicopter turn in the opposite direction. The small vertical propeller on the tail cancels this effect, producing an equal and opposite thrust.

What are helicopters normally used for?

Helicopters can reach places that no other vehicle, not even a plane, can reach (for example, mountaintops, small islands, unexplored forests, tops of burning buildings). Because of its special qualities, it is good for many purposes, other than transporting passengers and cargo. Examples are land surveying, rescues at sea or in difficult-to-reach places (see the examples on the left), laying cables, spreading insecticides, supplying isolated places, guarding and inspecting, firefighting and so on.

How much can a helicopter transport?

Cargo helicopters are large and have a considerable carrying capacity. For example, the Russian helicopter, MM-10 (shown above), has a fuselage almost one hundred feet long (33 m) and can carry up to fifteen tons on its cargo platform. The "sky crane" helicopters, used for unloading ships, can carry more than twenty-five tons.

SUPERSONIC JETS

What does "supersonic" mean?
"Supersonic" refers to airplanes that can fly faster than the speed of sound.

How fast does sound travel?
The speed of sound depends on the means. In water, for example it is 4,597 feet (1,400 m) per second, while in air, the situation that interests us, it is about 1,116 feet (340 m) per second, or about 720 miles (1,200 km) per hour.

When did the first supersonic flight take place?
The first supersonic flight was made on October 14, 1947 in a Bell X-1 experimental plane, piloted by an American Air Force captain, Charles Yeager.

What is the Concord?
It is the plane that you see in the photo below. A commercial airliner, it is the most famous and the fastest of the modern supersonic jets which fly at twice the speed of sound.

What is so special about the speed of sound?
There is nothing special about it: it is the speed at which sound travels. The reason for speaking about supersonic and subsonic (less than the speed of sound) aircraft has to do with the phenomenon that can be observed when a plane flies very quickly.

What is this phenomenon?
Sound is a wave-like phenomenon. It is made up of waves of compressed air which arrive at our ears and set our organs of hearing in vibration. When a plane flies at subsonic speeds, the sound it produces travels in all directions faster than the airplane itself. That is why we can hear an airplane arriving before it is in sight.

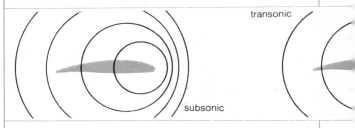

transonic

subsonic

What happens at a supersonic speed?
The sound produced travels at a speed less than that of the airplane (see the drawings at the top of the opposite page). Therefore, you no longer hear the sound of the plane approaching; you perceive it later. Exactly what you hear on earth is the "sonic boom," a kind of double explosion, which may be louder or softer depending on the size, speed and altitude of the plane.

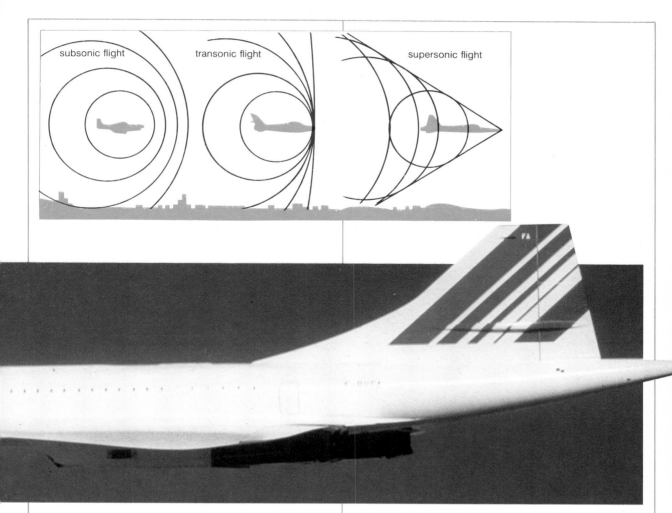

subsonic flight   transonic flight   supersonic flight

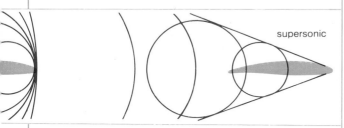

supersonic

What causes the sonic boom?

It is caused by two shock waves, some two to five hundred feet (60-150 m) apart, generated by a high-pressure cone extending backward from the airplane's nose and a low-pressure cone similarly projected from the tail.

What is transonic speed?

It is when a plane flies at the speed of sound. All the waves accumulate in front of the nose because they and the plane have the same speed.

What are the characteristics of a supersonic jet?

As the speed increases, the heat generated through friction increases. The supersonic jets must, therefore, have a much higher resistance and are usually made of stainless steel and titanium. The wings receive a lot of attention. In subsonic flights (see the drawings on the left), the lift is advanced; in the supersonic, retracted; and in the transonic, it moves back and forth. In this case, the plane experiences continual vibrations and must accelerate as much as possible to avoid excessive stress on its structure. In supersonic planes, the center of gravity must be moved to maintain a constant position without acting on the horizontal planes of the tail which would produce unbearable resistance. This can be done by moving ballast or fuel between the forward and rear tanks.

AUTOMATIC PILOTS

What is an automatic pilot?

The automatic pilot, or autopilot, is the group of devices which allow for automatic flight control. Originally, the main function of an autopilot was to maintain the direction and altitude determined by the pilot. Since a cruising airplane moves within large spaces, automatic control can be safely done, and the idea of an autopilot occurred early in the history of aviation. In 1914 in a flight over Paris, E.A. Sperry demonstrated the operation of the first automatic pilot.

How is the adjustment done?

Suppose that the nose of a airplane drops and the plane begins to lose altitude. This situation is sensed by a gyroscope (drawing at the bottom of the opposite page) which, preset to turn in a certain direction at a determined angulation, tends to return to the horizontal position in which it was set. It activates a small electric motor, the servo-motor, which moves the airplane's controls, making the nose climb. When the plane is again horizontal and flying at the preset altitude, the gyroscope, returns to its preset position, stops sending impulses and turns off.

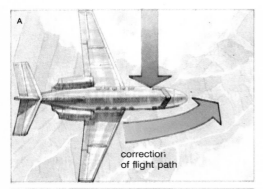

A

correction of flight path

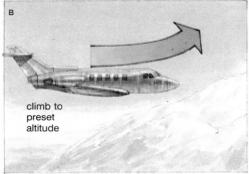

B

climb to preset altitude

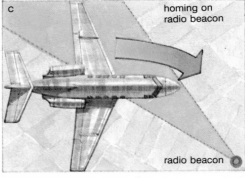

C

homing on radio beacon

radio beacon

D

semiautomatic landing

weather radar

How is an automatic pilot controlled?

The autopilot system is based on a device that compares the aircraft's movements (ahead, to the side, up or down) with the instructions introduced into the mechanism by the pilot. If, for example, the pilot has preset the autopilot for a horizontal flight at 9,850 feet (3,000 m) on a course of one hundred eighty degrees, any deviation from these instructions must be corrected by the autopilot as soon as it appears.

What if the plane deviates from the preset route?

The system is the same. In this case, a gyroscope which turns in a vertical plane, begins to work. Another servomotor activates the controls to bring the plane back to the original route. These devices can also activate the automatic pilot: accelerometer, wind-measuring devices (anemometers), barometers and special radio devices. The autopilot may control other electric, hydraulic or pneumatic servo-mechanisms.

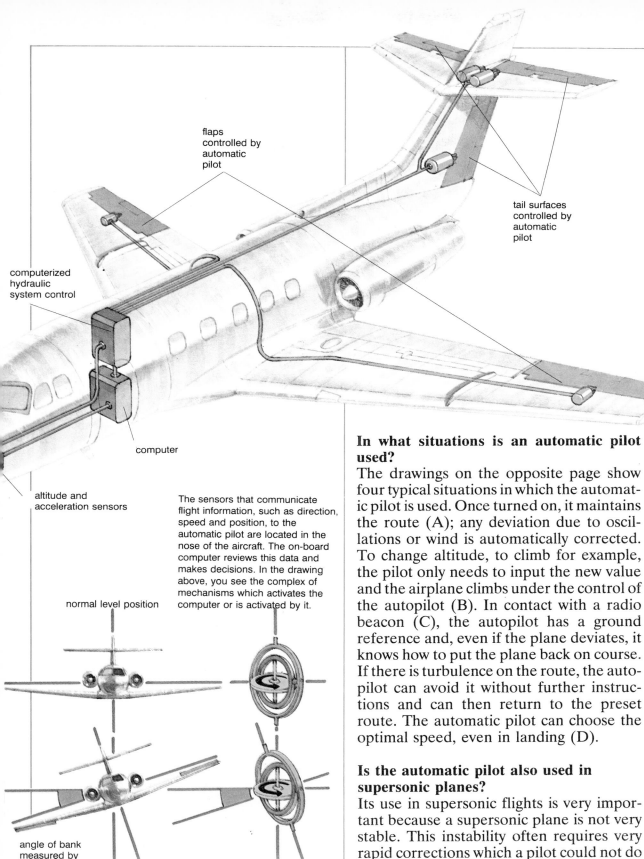

flaps
controlled by
automatic
pilot

tail surfaces
controlled by
automatic
pilot

computerized
hydraulic
system control

computer

altitude and
acceleration sensors

The sensors that communicate
flight information, such as direction,
speed and position, to the
automatic pilot are located in the
nose of the aircraft. The on-board
computer reviews this data and
makes decisions. In the drawing
above, you see the complex of
mechanisms which activates the
computer or is activated by it.

normal level position

angle of bank
measured by
gyroscope

In what situations is an automatic pilot used?

The drawings on the opposite page show four typical situations in which the automatic pilot is used. Once turned on, it maintains the route (A); any deviation due to oscillations or wind is automatically corrected. To change altitude, to climb for example, the pilot only needs to input the new value and the airplane climbs under the control of the autopilot (B). In contact with a radio beacon (C), the autopilot has a ground reference and, even if the plane deviates, it knows how to put the plane back on course. If there is turbulence on the route, the autopilot can avoid it without further instructions and can then return to the preset route. The automatic pilot can choose the optimal speed, even in landing (D).

Is the automatic pilot also used in supersonic planes?

Its use in supersonic flights is very important because a supersonic plane is not very stable. This instability often requires very rapid corrections which a pilot could not do manually.

179

ROCKETS

What is a rocket engine?
It could be defined as a sort of open furnace in which the fuel, either liquid or solid, burns very rapidly and the combustion products, the exhaust gases, escape quickly through an opening. The apparatus on which the rocket engine is installed is pushed forward for the same reason that a balloon spurts ahead when the opening is unblocked and the gas is allowed to escape.

What are the advantages of using rockets?
A rocket engine can work even without air because it carries with itself the fuel and the oxidizer needed for combustion. Oxidizers are substances, like oxygen, which make combustion possible.

Illustrated are some of the main space rockets: Vostok (1), is the Soviet rocket that launched the first astronaut, Yuri Gagarin, into space on April 12, 1961. The Atlas (2), launched the first American astronaut in 1962. Then in 1965, the Titan (3), launched Gemini 3, the first space capsule with two men aboard. The A-2 (4), is the rocket for the Soviet Soyuz spaceships. The Ariane (5), was built with the collaboration of several European countries. In 1969, the three-stage rocket Saturn V (6), launched the first men to land on the moon.

Are rockets jet-propelled?
Yes, they are. Rockets move as a reaction, going in the opposite direction to the exhaust gases coming out of the nozzles. This is another reason why rockets can move even in the absence of air, unlike planes and helicopters.

What fuels do they use?
Rockets for fireworks and for many ballistic missiles use solid fuels. Spatial launch rockets generally use liquid propellants, for example with hydrogen as the fuel and oxygen as the oxidizer. Nuclear rockets are still in the planning stages.

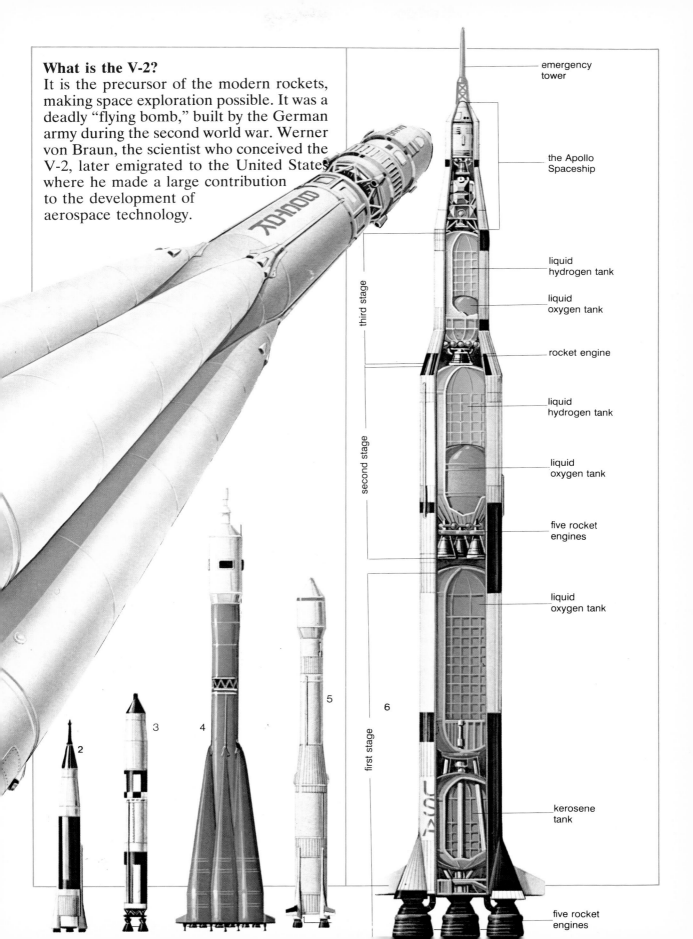

What is the V-2?

It is the precursor of the modern rockets, making space exploration possible. It was a deadly "flying bomb," built by the German army during the second world war. Werner von Braun, the scientist who conceived the V-2, later emigrated to the United States where he made a large contribution to the development of aerospace technology.

emergency tower

the Apollo Spaceship

liquid hydrogen tank

liquid oxygen tank

rocket engine

liquid hydrogen tank

liquid oxygen tank

five rocket engines

liquid oxygen tank

kerosene tank

five rocket engines

third stage

second stage

first stage

2

3

4

5

6

THE LUNAR EXCURSION MODULE

What is the LEM?

LEM stands for Lunar Excursion Module. It is the name of the Apollo system part that contained astronauts and was able to land on as well as lift off the moon.

What is the service module?

It is a large rocket that can be fired more than once, for example, to put the astronauts in orbit around the moon and then to leave that orbit and return to earth.

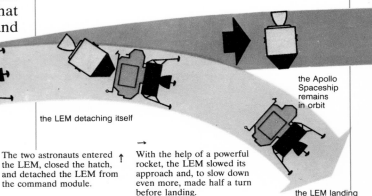

the Apollo Spaceship remains in orbit

the LEM detaching itself

the modules orbiting the moon

landing stage
lift-off stage
command module
service module

↑The three modules entered into lunar orbit. Two men moved from the control module to the LEM, ready to land on the moon.

The two astronauts entered ↑ the LEM, closed the hatch, and detached the LEM from the command module.

→
With the help of a powerful rocket, the LEM slowed its approach and, to slow down even more, made half a turn before landing.

the LEM landing on the moon

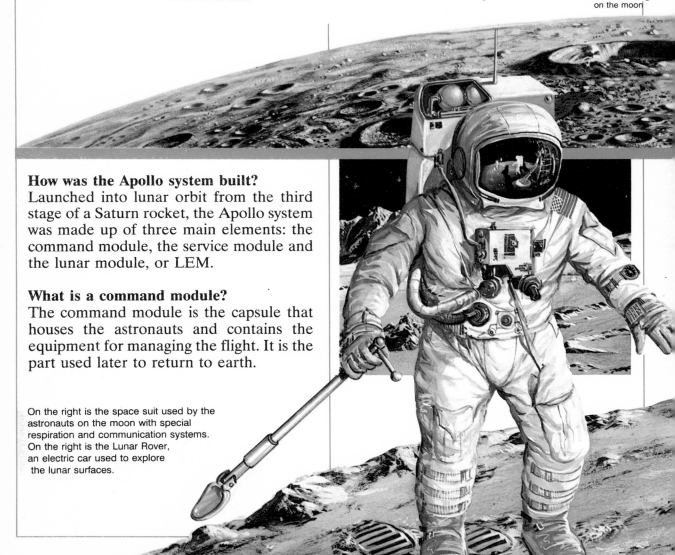

How was the Apollo system built?

Launched into lunar orbit from the third stage of a Saturn rocket, the Apollo system was made up of three main elements: the command module, the service module and the lunar module, or LEM.

What is a command module?

The command module is the capsule that houses the astronauts and contains the equipment for managing the flight. It is the part used later to return to earth.

On the right is the space suit used by the astronauts on the moon with special respiration and communication systems.
On the right is the Lunar Rover, an electric car used to explore the lunar surfaces.

Did the LEM return to earth?

No, it was abandoned on the lunar surface. Its structure would not bear the high temperatures produced by the friction with our atmosphere.

How did the moon landing occur?

You can see in the sequence of drawings below which show the LEM's landing and return to the orbiting spaceship. The lower part of the LEM, the landing stage, was used as a launching pad and then left on the surface of the moon.

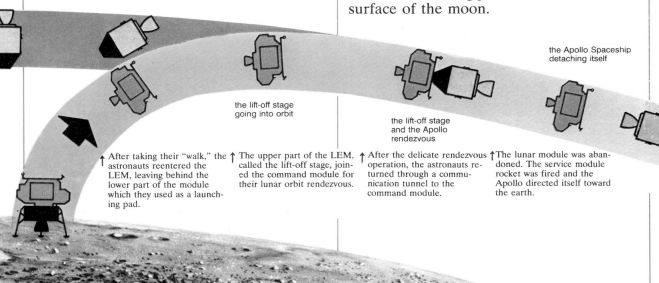

the Apollo Spaceship detaching itself

the lift-off stage going into orbit

the lift-off stage and the Apollo rendezvous

↑ After taking their "walk," the astronauts reentered the LEM, leaving behind the lower part of the module which they used as a launching pad.

↑ The upper part of the LEM, called the lift-off stage, joined the command module for their lunar orbit rendezvous.

↑ After the delicate rendezvous operation, the astronauts returned through a communication tunnel to the command module.

↑ The lunar module was abandoned. The service module rocket was fired and the Apollo directed itself toward the earth.

Why does the LEM have such a strange form?

There isn't any air in space where the LEM has to move and, therefore, aerodynamics are not important. So its designers could fully achieve the maximum utility, placing antennae and other instruments wherever they are needed.

How is the LEM made?

The LEM is a kind of mechanical monster, an extremely complex one. Nearly twenty-three feet (7 m) high, it is spiked with antennae for radio, television, radar and aptitude rockets. The upper part contains the cockpit and most of the antennae. Triangular portholes flank the entry, closed by a hatch. In the ceiling of the cockpit, there is another window for rendezvous and hook-up maneuvers between the lunar module and the Apollo Spacecraft.

What does the lower part contain?

It has four supports with telescopic shock absorbers and four "legs" with a large plate at the end of each one. The shock absorbers make landing on the surface of the moon even softer. This landing occurs at a very low speed, about 5.5 miles (9 km) per hour, due to the braking action of a powerful rocket.

ARTIFICIAL SATELLITES

What are artificial satellites used for?

They are vital components of modern international and domestic telecommunication systems. In addition, they are a basic navigational tool and they supply important meterological data. They are used to survey the earth's surface to provide information on the development and management of natural resources. They have been used for space and astronomical research and have assumed an important military role, especially in surveillance and reconnaissance.

How are artificial satellites put into orbit?

All satellites are put into orbit in basically the same way, by using multistage rockets that successively fall away as their fuel is used up. Once the satellite reaches an altitude of about two hundred miles (320 km), it is free of the drag of the earth's atmosphere and its movement is governed by the same laws that govern the orbits of natural satellites. A satellite rotating just above the earth's atmosphere completes an orbit in about ninety minutes. At an altitude of 22,300 miles (35,888 km), it takes exactly twenty-four hours. This orbit is called synchronous.

Some typical artificial satellites: the Space Shuttle (1); the orbital astronomical observatory, 1968 (2); LAGEOS (3), launched in 1976; TDRS (4), communication-relay satellite; marine-navigation satellite (5);

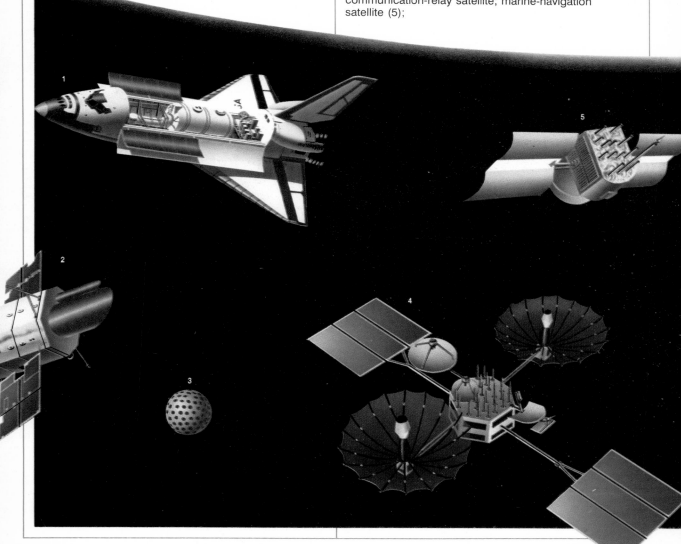

What is a geostationary orbit?

If a satellite moves in the same direction as the earth, from west to east, and if its orbit is over the equator, it will appear to be a fixed point in the sky. This is known as a geostationary orbit.

Is a satellite in geostationary orbit especially useful?

Yes, it is. Since it is constantly above the same area of the earth's surface, it is not necessary to have antennae on earth that must be continually reoriented to track the satellite and receive the signals it sends.

spatial telescope (6); upper atmosphere exploration satellite (7); Telstar (8), launched in 1962; Echo (9), low-pressure gas-inflated balloon of reflective material, launched in 1960.

Why are satellites so useful for telecommunications?

The microwaves used for telecommunications spread out in a straight trajectory and do not follow the curvature of the earth. Long-distance communication requires a series of repeaters placed close to one another (about thirty miles or 50 km apart) on elevated ground. These devices receive radio frequencies and retransmit them in a given direction. But this system can only work on land. A satellite, instead, can receive signals and retransmit them very much farther, even over water.

SPACE SHUTTLES

What are space shuttles?

Space shuttles are space vehicles that, unlike artificial satellites, can return to earth on their own, after having been in orbit for a given period. The Columbia, the first space shuttle (shown below), was launched in April, 1981.

How is a space shuttle made?

The main body, which resembles a small and chunky airliner, is the orbiter. It contains the working and housing space for a maximum of seven crew members and a cargo area fifteen feet (4.3 m) wide and sixty feet (18 m) long. At launching, this manned reusable orbiter is mounted piggyback on a large liquid-propellant tank and two solid-rocket boosters. These boosters are jettisoned when the shuttle reaches an altitude of about twenty-five miles (40 km), descending by parachutes to be recovered and reused.

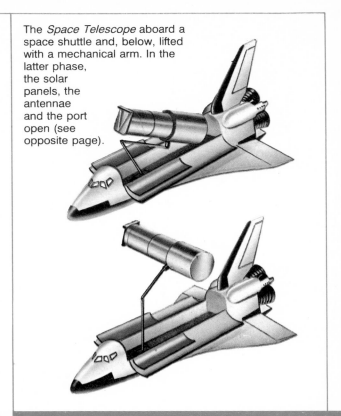

The *Space Telescope* aboard a space shuttle and, below, lifted with a mechanical arm. In the latter phase, the solar panels, the antennae and the port open (see opposite page).

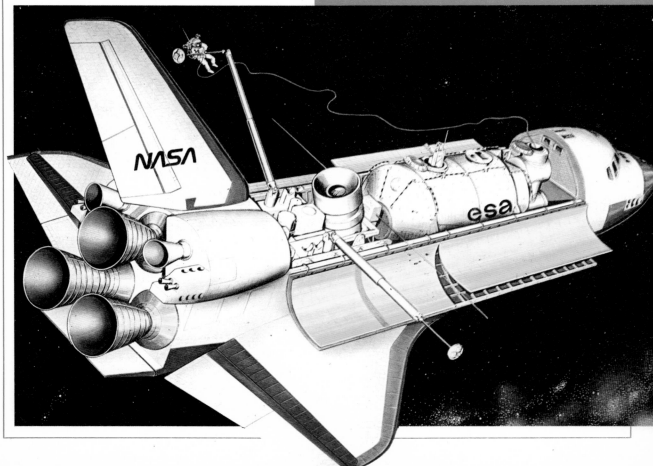

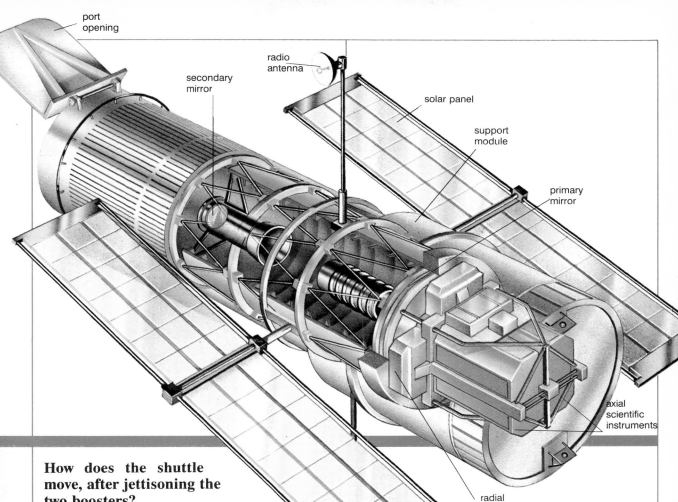

port opening

radio antenna

secondary mirror

solar panel

support module

primary mirror

axial scientific instruments

radial scientific instruments

How does the shuttle move, after jettisoning the two boosters?

At that point, two secondary engines, along with a group of smaller thrusters positioned along the body of the orbiter, push the spacecraft into orbit about 184 miles (295 km) high and enable it to maneuver in space.

How does the shuttle land?

The secondary and auxiliary engines enable it to leave its orbit. It can then sail to earth like a glider.

What can a shuttle be used for?

A shuttle is able to put telecommunication satellites in orbit. It could serve as a base for a telecommunication platform in space or for repair and maintainence of damaged satellites. It may also be used for military purposes, to put spy satellites or space control equipment into orbit.

What is the Spacelab?

The cargo area of a space shuttle can also carry the multi-purpose laboratory, called Spacelab, developed by the European Space Agency. (In the drawing on the opposite page, it is the somewhat cylindrical shape with the letters "esa" on it). Taking advantage of zero gravity and the near-perfect vacuum of space, the Spacelab can conduct scientific research in conditions that would be difficult to find on earth.

What is the Space Telescope?

The Space Telescope (see the drawings at the top of both of these pages) is a powerful forty-five foot (14 m) long space telescope which can be carried by a shuttle. It allows astronomers to see seven times deeper into space than they can from earth.

SPACE STATIONS

What is an orbiting station?

At present, it is a project for the future. It should, however, be a large station, permanently in orbit thousands of miles from the earth, away from the atmosphere of the earth. There the force of gravity should be almost totally absent.

What would it be used for?

Positioned outside of the earth's atmosphere with weak gravitational force, the station could be used for a large variety of experiments which cannot be done on earth. Some possible areas might be the study of astronomy, biology, materials and their manufacture and characteristics of the stratosphere. A completely new kind of industrial production might also begin in orbiting stations.

The illustration on the right shows a modular orbiting station made of elements transported by a shuttle. The size of the station may be increased by gradually adding new elements. The various parts are installed by mechanical "arms" which protrude both from the shuttle and from fixed points on the already-constructed structure. To the left of this drawing is a cross-section of an element intended to serve as a laboratory.

Has someone already planned the usage of an orbiting station?

Yes, various plans have already been proposed. One particularly interesting one was made by Werner von Braun (already mentioned as the designer of the V-2, the ancestor of modern rockets) in the early 1950s. He projected a ring-shaped station, placed in orbit at an altitude of 1,055 miles (1,700 km) with a crew of eighty people. Von Braun's idea has become well-known because it served as the model for the space station in the film, "Space Odyssey 2001."

Is the orbiting station in the picture a fantasy?

No, not exactly. It is one of the many models proposed by NASA, the American space agency. It gives an idea of how a station of this type might look. It shares a characteristic with all the other competing models: a modular structure. The various parts are prefabricated blocks, assembled, somewhat like pieces of Meccano, in orbit after being transported separately by one or more shuttles.

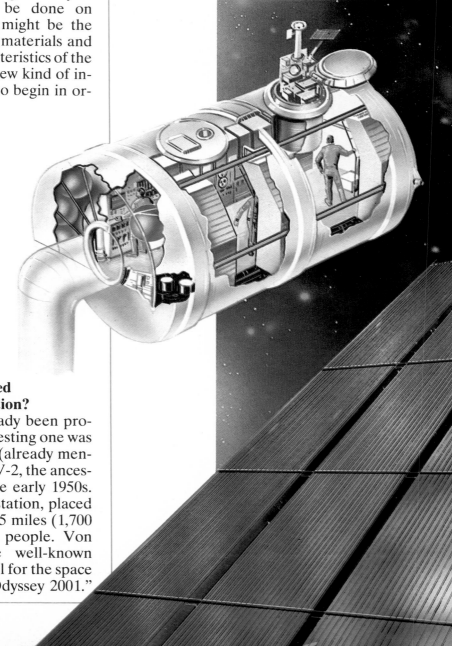

What elements does this station have?
There is a large central module which can be used to house the crew and the control and command equipment. Through various types of connections, this equipment manages the activities in the other parts of the station. Then there are lateral cylindrical modules which may be used as laboratories, shops and storerooms. Another important element is the large solar panels which guarantee the necessary supply of energy for the machinery and equipment of the station.

Why can new industrial production be started on space stations?
Gravity has a negative influence on many industrial processes which would have, instead, an ideal environment on a space station.

What are these processes?
Examples are the purification and solidification of substances, the separation of materials in solution (necessary for preparing many pharmaceutical products) and the manufacture of metal alloys and very pure varieties of glass.

DID YOU KNOW...?

What is a piston engine?

A piston engine is a hot-air engine in which an aeriform, such as water vapor or gas from burning fuel, pushes a piston back and forth within a cylinder. A crank transforms this lineal movement into rotary movement, for example, that of a wheel or a propeller. This engine is used a great deal. It is the engine found in automobiles.

What is a rotary engine?

It is also a hot-air engine, simpler than the piston type, in which the aeriform directly pushes the blades on the edge of a wheel, making them turn. The rotation may be transmitted to wheels, propellers or other turning parts. This type of engine is called a turbine.

What is the Otto engine?

This engine with the strange name is the typical car engine. Designed first by the German, N. Otto, in 1976, it is the well-known fourstroke combustion engine. In the engine, the fuel, gasoline, vaporized and mixed with air in the carburator, forms a wispy explosive spray which is injected into the cylinder. An electric spark, from a spark plug, provokes violent and rapid combustion of the gas-air mixture, creating high-pressure hot gases whose thrust makes the piston move.

What is a diesel engine?

The diesel is the strong workhorse of the internal combustion family. Rival to the internal combustion engine, which it resembles in structure and movement, it has no spark plugs. In the diesel engine, there are no little explosions. For this reason the engine is less subject to strain and uses a less expensive fuel. A diesel engine, however, must be more robust as it bears more pressure and needs an injection pump, which makes it more expensive to build.

Where are diesel engines used?

At first, they were mainly used in ships, in thermoelectric power stations and in heavy machinery. But now they are often used in automobiles as well.

What is engine output?

In every hot-air engine, heat "enters" and work "exits." Some heat, however, is also lost. A working engine is hot and heats up the surroundings, partly because the body of an engine, like a heater, gives off heat and partly because it gives off gas and hot steam. Heat given off in this way is wasted because its energy is not transformed into work. The less heat wasted, the more work the engine produces and the higher its output. For this reason, the output of a hot-air engine is the percentage of heat produced that the engine can effectively translate into work.

What is the typical output of the different kinds of hot-air engines?

The output of a steam piston engine is about 0.10, which means that about ten per cent of the heat produced is transformed into work while ninety per cent is dispersed. The average output for a steam turbine is about 0.30, for an Otto engine, about 0.25, and for a diesel engine or gas turbine, about 0.40.

What is the Skylab?

Skylab, literally meaning laboratory in the sky, was launched by Saturn V in the summer of 1973. It was a real sky station, weighing eighty-two tons and orbiting at 252 miles (420 km) around the earth. There was enough room in Skylab to allow people, doing various jobs, to remain for months. The astronauts reached Skylab aboard the Apollo capsule which, after a complicated approach maneuver, hooked itself up at one end of Skylab. Many important experiments were done in this orbiting laboratory. In the summer of 1978, Skylab began to leave its orbit and a year later, it fell to earth, luckily without causing any damage.

INDEX